AQA

information & communication technology

for AS Level

THIRD EDITION

Julian Mott ◄
Anne Leeming ◄

Edited by a Chief Examiner:
Helen Williams ◄

DYNAMIC LEARNING

Innovate • Motivate • Personalise

CD-ROM INSIDE

HODDER EDUCATION

NB Old Spec

The Publishers would like to thank the following for permission to reproduce copyright material:

Photo credits: p.29 *t* iStockphoto.com/Zolbayar Jambalsuren, *b* ©iStockphoto.com/216Photo; **p.30** *t*, **p.31** *t*, **p.32**, **p.100**, **p.147** and **p.209** Steve Connolly; **p.30** *c* ©iStockphoto.com/Oktay Ortakcioglu, *b* Getty Images; **p.31** *b* James Holmes/Science Photo Library; **p.33** *t* ©iStockphoto.com/Scott Vickers, *c* © Corbis, *b* © Corbis; **p.34** *t* © vario images GmbH & Co.KG/Alamy, *b* © kolvenbach/Alamy; **p.42** *t* © Corbis, *b* PurestockX; **p.43** *t* ©iStockphoto.com, *b* PurestockX; **p.44** © istockphoto.com/Valerie Loiseleux; **p.48** © Corbis, *b* © istockphoto.com; **p.49** © Helene Rogers/Alamy; **p.50** *t* ©iStockphoto.com/Dennys Bisogno, *b* © Niels-DK/Alamy; **p.51** © Nicola Armstrong/Alamy; **p.52** *t* ©iStockphoto.com/Aliaksandr Niavolin, *b* ©iStockphoto.com/Dmitriy Shironosov; **p.53** ©iStockphoto.com/Digital Planet Design/Sean Locke; **p.103** ©iStockphoto.com; **p.108** and **p.119** *l* Julian Mott; **p.119** *r* ©iStockphoto.com/Ross Elmi; **p.165** ©iStockphoto.com/Sandy Jones; **p.166** ©iStockphoto.com; **p.167**, **p.172**, **p.178**, **p.179**, **p.201** and **p.301** Anne Leeming; **p.168** © Blend Images/Alamy; **p.193** © Tony Charnock/Alamy; **p.205** Steve Chen/Corbis; **p.251** ©iStockphoto.com/Glenn Jenkinson; **p.270** © Najlah Feanny/Corbis; **p.272** © Blend Images/Alamy; **p.273** ©iStockphoto.com; **p.274** npower; **p.275** ©iStockphoto.com/Elena Elisseeva; **p.303** Fiona Hanson/PA Archive/PA Photos; **p.307** ©iStockphoto.com/Brad Killer.

Acknowledgements: p.197 Facebook; **p.219** Guardian Online; **p.227** Our Property.co.uk; **p.310** NHS website.
Every effort has been made to trace all copyright holders, but if any have been inadvertently overlooked the Publishers will be pleased to make the necessary arrangements at the first opportunity.

All exam questions reproduced with permission of the Assessment and Qualifications Alliance.

t = top, *b* = bottom, *l* = left, *r* = right, *c* = centre

Although every effort has been made to ensure that website addresses are correct at time of going to press, Hodder Education cannot be held responsible for the content of any website mentioned in this book. It is sometimes possible to find a relocated web page by typing in the address of the home page for a website in the URL window of your browser.

Hachette's policy is to use papers that are natural, renewable and recyclable products and made from wood grown in sustainable forests. The logging and manufacturing processes are expected to conform to the environmental regulations of the country of origin.

Orders: please contact Bookpoint Ltd, 130 Milton Park, Abingdon, Oxon OX14 4SB. Telephone: (44) 01235 827720. Fax: (44) 01235 400454. Lines are open 9.00 – 5.00, Monday to Saturday, with a 24-hour message answering service. Visit our website at www.hoddereducation.co.uk

© Julian Mott and Anne Leeming 2008
First published in 2008 by
Hodder Education
Part of Hachette Livre UK
338 Euston Road
London NW1 3BH

Impression number   5 4 3 2
Year   2012 2011 2010 2009 2008

Cover photo TEK Image/Science Photo Library
Typeset in Stone Informal 11pt by DC Graphic Design Limited, Swanley Village, Kent
Printed in Italy

A catalogue record for this title is available from the British Library

ISBN-13: 978 0340 958 308

Contents

Unit 1

Health and safety problems

People who use computers for long periods must use the equipment responsibly or they may face health problems which can be serious and long term.

Fortunately most of these problems are avoidable. You should be aware of what the problems are and how they can be avoided, not only for answering exam questions but also for your own health.

The main health problems associated with computers are discussed below.

▶ Repetitive strain injury

It is widely accepted that prolonged work on a computer can cause repetitive strain injury (RSI). The symptoms are stiffness, pain and swelling, particularly in the wrists but also in the shoulders and fingers. RSI can be a permanent injury which prevents the employee from working.

RSI occurs especially if users:

■ are carrying out repetitive tasks
■ have positioned the keyboard so that their hands and arms have to be held at an awkward angle
■ are squeezing the mouse too tightly.

A TUC (Trades Union Congress) report claims that young workers are more at risk from RSI than their older colleagues. The key factors that put them at risk include having to carry out repetitive tasks at speed, needing to use a good deal of force when working, not being able to choose or change the order of monotonous tasks and having to work in awkward positions.

The TUC data shows that 78% of younger workers have jobs which involve a repetition of the same sequence of movements and more than half the UK's four million workers aged 16–24 are forced to work in awkward or tiring positions. A young graphic designer took her employers, Shell UK, to court and won. She claimed that she was never shown how to use a computer mouse. She was awarded £25,000 damages for RSI that she began to suffer two years after joining the firm at the age of 20.

The risk of RSI can be reduced by:

- using a wrist rest while typing
- taking regular breaks away from the computer
- ensuring that computer users vary their work so that no user spends all of their working day at the keyboard
- having sufficient desk space to allow users to rest their wrists when not typing
- using specially designed "ergonomic" keyboards that force the user to adopt a more natural position of the arms
- ensuring that keyboards can be tilted and have well-sprung keys.

case study 1
▶ texting

In 2006, eight-year-old Isabelle Taylor from Lancashire developed repetitive strain injury (RSI) after sending up to 30 text messages a day. Isabelle had been using a mobile phone for two years. Then she started complaining about pains in her arms and hands.

RSI is normally associated with workers who use a computer keyboard for long periods. RSI can result from too much texting because mobile phone users tend to hold their shoulders and upper arms in a tense position.

Experts said that youngsters who overuse gadgets can suffer inflammation in the upper arms and wrists adding that the thumb is particularly susceptible to injury.

What advice would you give to Isabelle's mother to reduce the risk of damage to her daughter's health?

▶ Back problems

Sitting at a computer for a long period of time can lead to serious back problems. This is particularly true if you are in an awkward or uncomfortable position. The symptoms are back pain or stiffness, possibly a stiff neck and shoulders and sore ankles. Back pain from sitting at a computer is more common than back pain from heavy lifting!

Back pain occurs particularly if users:

- use computers for long periods
- adopt a bad posture – users should keep their backs straight and not slouch
- have their seat at an incorrect height.

Back pain is the largest cause of disability amongst workers in offices. The risk of back problems can be reduced by:

- having an ergonomically designed, adjustable swivel chair that supports the lower back and has a five-point base, which is more stable than just four legs
- adjusting the chair to the right height and the back of the chair to the right position to give back support

- adjusting the screen to the correct position using tilt and turn so that the user's back is not twisted or bent
- placing document holders next to the screen to reduce neck movement that can create back problems
- using a footrest
- getting up from the computer and taking regular breaks, say once an hour.

case study 2 ▶ back problems	Bad posture while sitting at a computer is more likely to lead to back pain than lifting and carrying heavy objects, say the British Chiropractic Association (BCA).
	56 per cent of BCA chiropractors said that office workers were more susceptible to back pain than workers with physical occupations.

▶ Eye strain

Looking at a computer screen for a long time can lead to eye strain. The symptoms can include headaches and sore eyes.

Eye strain occurs particularly if:

- lighting in the room is at the wrong level meaning that the screen cannot be read easily
- the screen is poorly sited causing glare or reflection on the screen
- the screen is of a poor quality and flickers
- the contrast on the screen is poor
- text is too small to read easily.

The risk of eye strain can be reduced by:

- having suitable lighting
- using non-flickering screens
- fitting screen filters to prevent glare and reflection
- fitting appropriate blinds (rather than curtains) to windows to reduce glare and prevent sunshine reflection
- if necessary, wearing appropriate spectacles while using a computer.
- ensuring that the eye line of the user is approximately level with the top of the screen and that the screen is slightly tilted
- regularly focusing on a distant object and then refocusing on the screen
- having regular eye tests (once a year) to ensure the use of appropriate spectacles.

▶ Epilepsy

It has long been suggested that flickering computer screens have contributed to the incidence of attacks in users who are epileptic.

The use of low-emission monitors and screen filters is likely to reduce the risk of triggering an epileptic attack in users.

| **case study 3** | In June 2007, an animation showing the logo of the London 2012 Olympics was removed from the organisers' website after reports that it had caused epileptic attacks. |
| ▶ **epilepsy fears** | The charity Epilepsy Action said it had received phone calls from eight people who had suffered fits after seeing the animation. |

▶ Stress

Nine out of 10 computer users in the UK say that they are regularly annoyed by their computers not working properly or running slowly and the time taken to fix problems. Stress-related sickness is estimated to cost the UK economy over £1 billion a year.

When computer frustration occurs, more than a third of IT users resort to extreme behaviour such as violence, swearing, shouting and frantically pressing random buttons.

The top five causes of stress for IT users are:

- slow performance and system crashes
- spam, scams and too much email
- pop-up ads
- viruses
- lost or deleted files.

The best thing to do is to calm down, breathe deeply and remind yourself to keep your cool according to Mike Fisher from the British Association of Anger Management.

Other recommended ways of reducing stress include:

- not trying to save money when purchasing equipment
- installing hardware that is capable of meeting the demands of the tasks
- choosing well-designed software that has an interface that is appropriate for the user
- having a reliable IT support team
- providing users with adequate support and training
- saving your work often as you go along
- backing up regularly so that you don't lose your work
- using spam filters and anti-phishing and anti-virus software
- going for a short walk to get some fresh air
- good ergonomics: make sure that your position at the computer is comfortable
- removing the computer from your bedroom: take the stress of your work away from where you should be relaxing.

▶ How you can reduce the risk of health problems

During this course, you will be using computer equipment for long periods of time, particularly when you tackle practical tasks. Unless you take the right actions, you could put your health at risk.

You should:

- not work continuously at a computer for long periods
- get up and walk around occasionally (not all the time!)
- focus on a distant object every so often and then back on the screen
- ensure that you have a work area that is ergonomically suitable: not too cramped, with suitable seating and with the computer and the screen at the right height.

Health and safety legislation ◀

The law on the use of ICT equipment at work is covered by the *Health and Safety (Display Screen Equipment) Regulations* of 1992. This law says that employers must take various measures to protect the health of workers using computers.
The law says that employers must:

- provide adjustable chairs
- provide screens that can be tilted
- provide anti-glare screen filters
- ensure workstations are not cramped
- ensure room lighting is suitable
- plan work at a computer so that there are breaks
- provide information on health hazards and training for computer users
- on request, arrange and pay for appropriate eye and eyesight tests for computer users
- provide special spectacles if they are needed and normal ones cannot be used.

Note: These regulations apply only to staff in offices and not to students in schools or colleges.

Health and safety and the design of new software ◀

In designing new software, designers must be aware of the risks to the users' health. Their designs should minimise health risks.

■ Screen layouts should be clear so that eye strain is minimised.

■ Colour schemes should not be too bright and should have good contrast otherwise eye strain could occur.

■ Text should be a suitable size and in a clear font, minimising the chance of eye strain.

■ Instructions and help facilities should enable learning, otherwise the user could feel stressed. Clear and helpful error messages allow users to put their mistakes right for themselves.

■ Menu systems should be well structured and include short cuts so that the number of keystrokes or mouse clicks is minimised. Lengthy navigation can be stressful.

■ Annoying sounds and flashing images should be used sparingly to reduce the possibility of an epileptic attack and reduce eye strain and stress.

■ Software should be compatible with other packages otherwise slow use could lead to stress.

■ The use of drop-down lists can reduce data entry and help prevent RSI.

■ Logically ordered fields should be used to make data entry easier and reduce stress; the fields on the screen should be in the same order as on a written document from which they are being copied as it can be annoying and time consuming to have to scan the written document for the required data.

<tip>

You need to take all these features into account when designing screens for your solution in Chapter 3. Many students fall into the trap of designing very busy screens with very brightly coloured backgrounds and a mixture of fonts.

▶ Worked exam question

Describe three features of poorly designed software that can cause stress or other health problems to a user. (6)

January 2001 ICT1

▶ EXAMINER'S GUIDANCE

As this question allocates 6 marks for three features you need to gain two marks for each feature, so just listing three features is not enough.

You need to identify a feature and then explain why it would cause stress or other health problems.

There are many features that you could choose from; don't choose the first ones that come into your head but rather choose features about which you can write enough!

The information in the preceding section itemises features that developers should include for good software design; this question is asking for the opposite, so you need to turn things around a bit. One point made is that "colour schemes should not be too bright and should have good contrast otherwise eye strain could occur"; this could be turned around to:

▶ SAMPLE ANSWER Screen designs that have very brightly coloured backgrounds with text displayed in a poorly contrasting colour can cause eye strain.

Now add three examples of your own.

Activity 1

Find out more about health hazards and ICT at the Health and Safety Executive's web site: http://www.hse.gov.uk/. List the services that the site provides for a) an employer b) an employee.

The Trades Union Congress (TUC) health and safety newsletter site, http://www.hazards.org/ is worth exploring too.

Activity 2

Produce an induction booklet for new employees that provides them with guidelines in how to use the computer equipment safely. You should include a description of the organisation's responsibilities to them.

Activity 3

Copy and complete the table below (some ideas have been included).

	Associated health issues	Prevention steps that can be taken – there may be more than one answer
Keyboard		Tiltable keyboard
Chair		
Software packages		
Computer desk and surroundings		
Office and surroundings		
Monitor	Glare can cause eye strain	
Printer		

SUMMARY

Regular use of **ICT** equipment over a long period of time may lead to health problems, particularly:

▶ **RSI (repetitive strain injury), often stiffness and swelling in the wrists**

▶ **eye strain**

▶ **back problems.**

Incidence of these problems can be reduced by taking sensible precautions such as not using equipment for too long, introducing adjustable chairs and using wrist supports and screen filters.

New regulations have been introduced throughout the European Union, setting standards for using **ICT** equipment in offices. Employers have duties to:

▶ **provide adjustable chairs**

▶ **provide adjustable screens**

▶ **provide anti-glare screen filters**

▶ **pay for appropriate eye and eyesight tests**

▶ **plan work so that there are breaks from the screen**

▶ **provide information and training for ICT users.**

Software should be designed to minimise health risks by having, for example:

▶ **clear screen layouts**

▶ **muted colour schemes**

▶ **readable text**

▶ **help facilities and error messages**

▶ **keyboard short cuts**

▶ **no annoying sounds or flashing images.**

Questions

1. Your sister has just left university and has got a job creating websites and animations. This means that she will be sitting at a computer for most of her working day. What advice would you give her about protecting her health while at work? (3)

2. If software is not designed properly, it can cause stress in a user. For example, if a data entry screen is cluttered with multi-coloured and unnecessary images, the user can become confused and disoriented.
Describe three other features in the design of a software package that could cause stress in a user. (6)

3. An ICT professional within a company has been asked to produce health and safety guidelines for employees working with ICT. The guidelines will be stored on the company's intranet so that all employees can access them at all times.
State four guidelines that you would include and explain the reason for each. (8)

4. Mr Hadawi is setting up a new office for his car-hire business. He needs to equip the office with a computer that will be used all day by his employee. Mr Hadawi wishes to ensure the health and safety of his employee.
Describe two features that Mr Hadawi should consider when buying and installing each of the following:
   a) the screen
   b) the chair
   c) the keyboard
   d) the desk and surroundings. (8)

5. A clerk working in the accounts department of a large company spends all day entering employee timesheet data into the company's payroll system. The clerk uses a terminal linked to the company's main computer.
To ensure the health and safety of the clerk, state, with reasons:
   a) two work practice procedures that the company could introduce (4)
   b) two design features that the workstation the clerk uses should have (4)
   c) two design features that the software the clerk uses should have. (4)

   June 2005 ICT1

6. Read case study 4. Describe four actions that Andrea Osbourne could have taken to prevent the injury occurring. (8)

case study 4
► RSI damages

In 2006, a night editor who was eventually forced to leave a newspaper after developing RSI was awarded £37,500.

Andrea Osbourne worked at the paper for two and a half years, using a mouse for an average nine hours per night and up to 45 hours per week, without a break.

She was told by a hospital consultant that she would never be able to do that type of work again after she developed stiffness and pain in her right elbow and she was advised to seek an alternative career.

2 Analysis of problems and software

AQA Unit 1 Section 2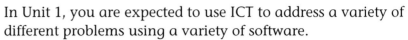

In Unit 1, you are expected to use ICT to address a variety of different problems using a variety of software.

As well as creating the solution, you need to produce the following documentation:

- analysis of the problem – problem identification, client requirements and lists of inputs, processing and outputs
- test documentation including a test plan and test evidence.

This documentation is called the **sample work**. It should be word-processed and include screenshots and printouts from the solution that you have created. It should be between 10 and 20 pages long.

You will take the sample work into the Unit 1 examination with you and use it to help answer some of the questions on the exam paper. The sample work will be handed in with your question paper or answer booklet at the end of the exam.

Pages of the sample work should be numbered and it is good practice to put your name, candidate number, centre name and centre number in the header or footer at the top or bottom of each page.

For the exam you need to know how to design and evaluate your solution.

Choosing the problem

▶ What software can I use?

The software that you use will depend on the problem that requires solving. You need to provide evidence of processing text, images, numbers and sound.

The following types of software are all likely to be suitable:

- animation software, such as Macromedia Flash
- desktop-publishing software, such as Microsoft Publisher
- image manipulation (graphics) software, such as Adobe Photoshop
- presentation software, such as Microsoft PowerPoint
- relational database management software, such as Microsoft Access
- sound-recording and -editing software
- spreadsheet software, such as Microsoft Excel

- web development software, such as Macromedia DreamWeaver
- word-processing software, such as Microsoft Word.

\<hints\>

- You don't need to use all these software packages. In fact you wouldn't have time to tackle every one of these in the appropriate detail.

- Remember that you need to process sounds so you are likely to need to include at least one of the following: animation software, presentation software, sound-recording and -editing software or web development software.

- A problem may require the use of more than one piece of software.

▶ What sort of problems should I tackle?

The examination board gives some sample problems. They are:

1. an electronic photo album
2. a rolling multimedia presentation for a school open day
3. a website for a local nursery
4. organising a blog for a local councillor
5. producing invoices for a small business
6. organising a set of podcasts for a teacher
7. an interactive multimedia display for a tourist information centre.

Possible software to solve these problems might be:

1. A scanner and appropriate software may be needed to input images. Images may be manipulated using graphics software. If the images are to be viewed over the Internet, web development software may be needed. Presentation software could be used instead to solve the problem.
2. Presentation software is needed but graphics software and sound-editing software may also be needed.
3. Web development software, graphics software and sound-editing software.
4. Web development software.
5. Spreadsheet software.
6. The podcasts may need to be recorded and edited using sound-manipulation software. They may be stored on a computer and accessed using hyperlinks on web pages set up with web development software.
7. This could be set up using animation or presentation software.

This demonstrates the wide variety of problems to be solved. However there are many similarities:

- they are all problems that can be solved using ICT
- each problem needs to be analysed to keep a record of exactly what is required
- each solution needs to be designed, implemented, tested, evaluated.

You are strongly advised to tackle at least two tasks. It is a good idea to take into the exam one piece of work that involves presenting information, such as **creating a website**, **a presentation** or **a desktop-published document**, and one piece of work that involves calculations and processing, such as **a spreadsheet**.

In this book, we look at how to carry out two types of task:

- creating a solution based on a website
- creating a solution based on a spreadsheet.

Creating a website ◄

▶ Problem identification

You need to start with a description of a problem that requires solving, stating:

- who is the client
- who is the user
- who are the audience (Remember: Not all problems will have an audience.)
- a brief description of the problem that needs solving.

example	The client is Jodie Watson, who owns a small company letting holiday cottages in France, Italy, Spain and Germany.
▶ **initial problem description**	Her company, Perfect Cottages, employs two other people in her office in Derbyshire. Jodie would like a website to:

- display photos of the holiday cottages

- display the prices

- display maps showing where the cottages are

- give dates of forthcoming shows at which she will have a stall, where customers can talk to her about the sort of cottage that they want

- enable potential customers to contact her.

She has asked me to help.

▶ Finding a client

Ideally your client will be a real person and the problem will be a real problem. However this is not always possible. If you cannot find a client:

- Don't pretend that someone famous like Wayne Rooney or Arnold Schwarzenegger has asked you to solve a problem for them.
- Don't pretend that a large company like Ferrari or Dolce and Gabbana have invited you to help them.
- Do get an adult such as your teacher or a parent to act as if they are the client.

▶ Interviewing the client

You need to interview your client to find out exactly what they want from the solution you have to create. You do not have to interview them face-to-face. If they are some distance away, you can do this by phone, email or instant messaging.

- Who will the user be? The user in this case is not someone who visits the website – they are the audience. The user is the person who will update the website. Have they any experience of this sort of work? The user may be the same person as your client.
- What will the output be? If you know what the output needs to be, you can work backwards to find out what inputs and processing are required.

It is a good idea to write down all the questions that you want to ask the client before starting the interview. This will ensure that you don't miss anything out or waste their time while you think of questions and write them down.

---

example
▶ **questions for the client**

Questions about the user:

- Who will keep the site up to date?
- Have they ever updated a website before?
- Are they computer literate?
- Are they logical and organised?
- How often will the site be updated?

Questions about the output:

- Do you want a home page and several other pages?
- What images do you think you want on the site?
- How many are there likely to be?

example (contd)

- How big are these images?

- Is there text to go with the images?

- Who will provide the text for the images and other wording for the site?

- How many forthcoming show dates are there?

- What contact details do you want on the site?

- Does the company have a logo?

- If yes, do you have an electronic version of it?

- What are the company colours?

- Is there a company font?

- Who is the intended audience for this website?

- What image do you want the site to have? Reliable? Conservative? Modern?

Once you have written down the questions, you are ready to interview the client. It is often a good idea to keep a verbatim record of the answers.

example
▶ record of interview with the client

Interview with Jodie. 29 September.
Me: Who will keep the site up to date?
Jodie: My assistant, Monica Holden. She deals with all the computer stuff. I'm hopeless with computers but she's very good.
Me: Has she ever updated a website before?
Jodie: I don't think so.
Me: I will provide full instructions on how to update the site. It really is very easy. She is computer literate?
Jodie: Oh yes. She deals with all the emails and so on.
Me: Is she logical and organised?
Jodie: Oh yes. She's very reliable and she needs to be. An out-of-date website is worse than no website.
Me: How often will the site need to be updated?
Jodie: It must be updated after every show. It's no good advertising a show that was last week.
Me: And how often are the shows?
Jodie: Once or twice a month throughout the winter. Not many in the summer. But the website will need to be updated straight after the show. The next day at the very latest.
Me: Will there be any other updates? Do your prices change?

Jodie: **Sometimes the prices go up. Sometimes we might cut prices or do two weeks for the price of one. We need to be able to add new cottages and delete old ones, of course.**

Me: Do you want a home page and several other pages?

Jodie: **Well we have cottages in four countries: France, Italy, Spain and Germany. On the first page, I want four buttons; one to take you to the list of all the French cottages, one to take you to a list of the Italian cottages, and so on.**

Me: And I suppose you'd like buttons on these pages to take you back to the home page.

Jodie: **Yes, please. And I'd like something about the company.**

Me: So on the first page you will have details of the company and links to the four countries. You could have the name of the country and the country's flag as the link. How does that sound?

Jodie: **That sounds brilliant.**

Me: So we'd need images of the four flags. How many other images do you think there are likely to be on the site?

Jodie: **Well we have 65 cottages in France. I think there are 34 in Spain, 23 in Italy and 19 in Germany. I want at least one large photo of each property. I've seen websites where they have a list of properties with a small photo. If you click on this photo, you then see bigger photos. Can you do that?**

Me: Of course. The small photos are called thumbnails. They are hyperlinks to other pages. That way the pages load more quickly. How big must the bigger pictures be?

Jodie: **It's important to see the features of the cottage. Perhaps half a screen. And there has to be a map. That has to be clear enough to read.**

Me: Is there text to go with the photos?

Jodie: **Yes. There's quite a lot actually, such as the cottage code, the cottage name, the village, the region, the price for one or two weeks, high or low season. I can give you examples of all the data, if you like.**

Me: That would be very helpful. Can you email it to me?

Jodie: **Of course, but it's quite long for each cottage. Will you be able to fit it all in with the large photo?**

Me: That shouldn't be a problem. I am a bit worried about the France page. If you have thumbnails of all 65 cottages on one page, it will not be very user friendly. We could break it up into three or four pages – perhaps by location in France. Say the west, the south, the north and the east.

Jodie: **Maybe. Or by how many the cottage sleeps. Let me think about that. I'd like to see all 65 cottages on one page to start with. If it doesn't look right, you can easily change it, can't you?**

Me: Of course. Do you have suitable photos already?

Jodie: **Yes, but not on computer.**

example (contd)

Me: I can scan them in. Would you like mouseovers?

Jodie: What are they?

Me: As you move the mouse over a hyperlink it is highlighted in some way. The text might change colour or go bold.

Jodie: Oh yes, I've seen them. They are a good idea.

Me: Who will provide the text for the site?

Jodie: Me. It will all be typed up. It will be in Word format. Is that OK? I can email it all to you.

Me: That's good. How many dates of forthcoming shows are there?

Jodie: I already have eight dates pencilled in for the autumn. They will pick up again soon.

Me: What contact details do you want on the site?

Jodie: I will supply that too but it's the address, the phone and fax numbers and the email address.

Me: Does Perfect Cottages have a logo?

Jodie: Yes. Haven't you seen it on our brochure? Oh, I'd like it on every web page.

Me: Do you have an electronic version of it?

Jodie: Yes, I can supply it.

Me: What are the company colours?

Jodie: Navy blue and red. But the logo has a pale blue background.

Me: Red or navy blue will not look good as the background to a web page but a light blue background with little flashes of red and navy blue would look good. Would that be OK?

Jodie: Yes, but I'd like to see the designs first.

Me: Of course. Is there a company font?

Jodie: No. Can you suggest one?

Me: Yes I will. Who is the intended audience for this website?

Jodie: Well, our customers are mainly families of course; the parents are mostly in the 30 to 45 age range. They will be experienced users of the web. The pages must be in a logical order. What do you call it when the order of the links is obvious, even if you use the site for the first time?

Me: Intuitive?

Jodie: That's it. Intuitive.

Me: What image do you want the site to have? Conservative? Modern?

Jodie: Modern, please. Perfect Cottages is anything but staid. We are a go-ahead company providing the best in holidays.

Me: Thanks, Jodie. I'll produce a list of your requirements and get back to you. Later I'll do some rough designs.

▶ Other ways of finding out information

There are other ways of finding out about the problem other than interviewing the client. If you have a real problem you can:

- Inspect documents
- Observe the current system (if there is one) in action
- Send questionnaires to users.

The data that Jodie sent on one of the cottages was as follows:

example
▶ follow-up email

From: Jodie Watson
To: Katie Gibson
Date: 30 September
Subject: Sample data

Katie

Here is the sample data on one cottage that I promised you.

Cottage Code:	F0013
Cottage name:	Le Vieux Moulin
Village:	St Georges
Region:	Normandy
Map reference:	C3
No of bedrooms:	3
Number of beds:	6
Changeover day:	Saturday

Prices:

Low season, one week:	£436
Low season, two weeks:	£708
High season, one week:	£679
High season, two weeks:	£1198

Facilities:

Swimming pool:	No
Garden:	Yes
Microwave:	Yes
Dishwasher:	Yes
Washing machine:	Yes
TV:	Yes
Bed linen for hire:	Yes

Charming, converted old mill dating from 1816, eight miles from the town of Avranches. Sailing, fishing, woodland walks and golf nearby. The port of Caen is one hour away and Cherbourg two hours.

Local attractions:

The tapestry at Bayeux; cathedrals at Coutances, Lisieux and Rouen; Mont St Michel; the D Day beaches; World War II museum at Avranches; the memorial museum at Caen.

▶ Defining the client's requirements

You now need to produce a list of the client's requirements. You should show them to your client and get their agreement.

If not, it can lead to a dispute later on, so write down a full list in as much detail as possible.

example
▶ **client requirements**

Perfect Cottages website – problem description
Problem: Jodie Watson (owner of the company) requires a website to tell people about her holiday cottages.
Client: Jodie Watson
User: Monica Holden
Audience: Mainly parents in the 30 to 45 age range.
Perfect Cottages website – client requirements

1. Easy to update

2. Intuitive for the audience

3. Quick to load

4. Modern image

5. Company logo on each page

6. Verdana font – I discovered this was the one used in Perfect Cottages' logo.

7. Pale blue background but with navy blue and red flashes; text navy blue or red – I will use a cascading style sheet to set the colours.

8. Links to other pages and photos

9. Thumbnail images as links

10. Mouseovers for hyperlinks

11. Thumbnails and photos of all cottages (141 in total)

12. Details of all the cottages such as the cottage code, the cottage name, the village, the region, the price for one or two weeks, high or low season

13. A page for each cottage with photos, taking up about half a page, and a map

14. Contact details

15. Details of the company

16. Full updating instructions

17. Ability to update prices, delete cottages, add new cottages

18. 100% working in different browsers and with different resolutions

▶ **Defining the inputs, processing and outputs**

Once you have agreed the requirements with the user, you need to start thinking about the website. Start with the outputs

– what the solution will tell users or the audience – and work backwards.

Outputs

Make a list of the outputs. Our example is a relatively small website and there are not many outputs.

---

example

▶ **outputs**

The outputs will be intuitive web pages containing:

1. A home page linked to pages for each country

2. List and flags of the four countries

3. Thumbnail images of the cottages

4. Maps

5. Links to a page for each cottage

6. Images of the cottages

7. Information about the cottages – the cottage code, the cottage name, the village, the region, the price for one or two weeks, high or low season and other important details

8. Details of forthcoming shows including dates and the venue

9. Contact details

10. Details of the company

11. The company name and logo

12. Links to other pages

Inputs

Make a list of the inputs. What data is needed to produce the required outputs? In this case, the inputs are similar to the outputs.

---

example

▶ **inputs**

The inputs will be:

■ Images of 141 cottages – to be scanned in

■ Raw text (about the 141 cottages) – to be provided by Jodie

■ Raw text details of forthcoming shows – to be provided by Jodie

■ Raw text contact details – to be provided by Jodie

■ Company logo – to be provided by Jodie

Processing

What processing is involved in converting the data inputs to information output?

example
▶ **processing**

The processing required will be:

- Resizing the images of the cottages to fit on the screen
- Cropping the images as required
- Resizing the images of the cottages to create the thumbnail images
- Cropping the thumbnails so that they are roughly 100 pixels square

Creating a spreadsheet solution ◀

If you are creating a spreadsheet solution, you still need to list inputs, processing and outputs. Suppose you are producing a quotation system for a small printing company.

example
▶ **inputs**

Inputs will include items such as

- the type of paper or card
- the quality of the paper or card
- the quantity required
- whether it is folded or not

example
▶ **processing**

The processing will include:

- looking up prices
- calculating the totals
- adding on VAT
- filing details in another sheet

example
▶ **outputs**

The outputs will include a quotation worksheet containing:

- the company name and address
- the company logo
- the paper or card selected
- the number required
- whether it is folded or not

example (contd)

- the quality of the paper
- the total price
- the date
- error messages.

Questions

Choose one of the following scenarios:

- Uncle Horace wants to store his digital holiday photos in a format in which he can easily see them.
- A local hot-food, take-away restaurant requires a website to display prices, photos, opening times, delivery costs and how to get to the restaurant.
- The captain of Rottingdale Cricket Club wants a website to show visitors results of matches this season and forthcoming fixtures. They have two teams, the first XI and the second XI and they play matches every Saturday and Sunday throughout the summer.

- The Bridge Hotel near Snowdon in North Wales would like a small website to advertise their 12 single and 23 double rooms.
- The West Gower Youth Centre wants to set up a website so that youngsters can find out what is happening at the centre in the next month.

1. Write down at least 12 questions to ask the client.
2. Practise by asking someone in your class and recording their answers.
3. Write down the outputs.
4. Write down the inputs.
5. Write down the processing.

3 Design of solutions

Once you have completed the analysis and described the inputs, processing and outputs, it is tempting to go straight to the computer to implement your system.

However it is essential that you do the design work first away from the computer, otherwise you will find yourself doing more work in the long run. By producing designs you can:

- show the designs to the client to get their comments and approval
- ensure that all the input, processing and outputs are included
- ensure that your designs are aesthetically pleasing
- include appropriate validation techniques.

Design obviously includes screen designs but other techniques will depend upon the solution chosen. They may include printed outputs, record structures, data-capture forms, data-entry methods, audio capture, image capture, video capture, validation techniques, animations, macro designs and security methods.

When your designs have been completed, you should be able to give them to a third party (such as someone else in your class) and they should be able to implement the solution. This is good practice. In a commercial environment if a developer leaves a company for some reason, whoever takes their place should not have to start again but should be able to develop the solution from the original designs.

Professional examples

Before producing your design look at professionally produced examples. (Some courses insist that you do this.) Study genuine websites on the Internet. What colours do they use? What fonts and font sizes do they use?

For example, many successful commercial websites:

- use white or a pale colour for the background
- have small images
- put text in three or four columns
- use a clear, clean font such as Verdana

- use size 8 in a dark colour for normal text
- use dark backgrounds sparingly.

Choice of software ◀

You need to specify the program or programs that you will use to implement the solution, as this may affect the designs. There is more on software selection in Chapter 7.

Designing a website solution ◀

You need two sorts of design. The website structure diagram shows how the pages link together. The other designs show what each page will look like.

▶ **Website structure diagram**

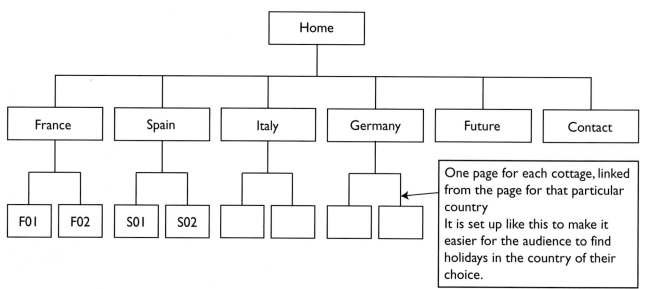

Figure 3.1 Structure diagram for Perfect Cottages

- Do I need to include every single page in this diagram?
  No. This is just to give an indication of the links in the site. You could not fit in all 141 pages about the cottages.
- Should I annotate the diagram?
  Yes, if you think you need to add extra information to make the design clear.
- Is it necessary to have a structure like the one above? Couldn't I just link from the home page to all the other pages?
  That will depend on how many pages there are on your site. As there are 141 cottages, providing links to a different page for each cottage would require 141 hyperlinks on one page. This would be difficult to fit in.

▶ Page design

It is a good idea to take a screenshot of a blank page in a web browser. Print it out and use it as the basis for your design like the one in Figure 3.2.

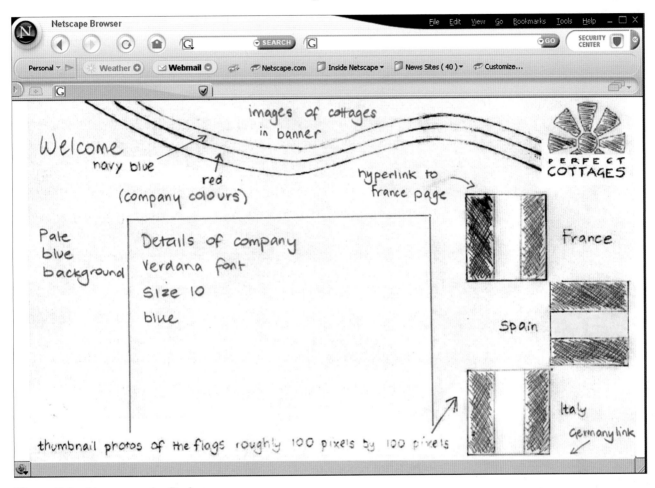

Figure 3.2 Home page for Perfect Cottages

- At this stage, designs are only rough but that does not mean that they are untidy. Scribbled designs on screwed-up paper will impress nobody. However you are not being marked on your drawing skills.

- Annotate your screen designs as in Figure 3.2 to show features such as hyperlinks, mouseovers, etc.

- Do I need a design for every page?
  Not if pages are similar. For example, the page for Italy will look very similar to the page for Spain. In this example, it would be sufficient to produce:
  - *a design for the home page*
  - *a design for one of the country pages, e.g. Spain*
  - *a design for one of the cottage pages, e.g. Le Vieux Moulin*
  - *a structure diagram for the website.*

- Should I create my designs by hand or on computer?
  Most students find it quicker and easier to draw their designs by hand. If you have difficulty producing hand-drawn designs that aren't scruffy, by all means produce the designs on computer but be prepared to spend more time on them.
- Should I use a pencil and a ruler?
  Of course.
- How big should the designs be?
  Don't make your designs too small. Make sure any writing on the design is clear and big enough to read easily. Take a whole sheet of A4 to display the design of one page.
- Can I create my designs in web creation software such as DreamWeaver, FrontPage or Expression Web?
  No. Designs should not be in the software that you will use to create the solution.
- Should I do my designs in black and white or in colour?
  Either black and white or colour is fine. If your design is black and white and an area of the screen is a certain colour, label it to say so. Once you have shown these rough designs to your client and got their approval, you can produce a full-colour design if you want.

▶ Printed output

Don't just think about screen designs. Will the users or the audience need to print out web pages? What will the printouts look like? Do you need a printer-friendly version of a web page?

Produce some mock-ups to show your client. Specify the fonts and the font size. If the background is in colour, state what the colour is.

Make sure that the output content matches the paper size. The paper size does not have to be A4 and the colour of the paper may not always be white.

▶ Security methods

You also need to mention security. Is a password required to edit pages or to upload them to the Internet? Do you want to create a backup every time you save? If this is important you need to define it in this section.

Designing a spreadsheet solution ◀

If you are designing a spreadsheet, you should use similar techniques.

Look at professional spreadsheets. For example, if you are creating a system to produce an invoice, study professional invoices. What information is printed on the invoice? What size is the text? What fonts and colours are used?

It is a good idea to take a screenshot of a blank spreadsheet. Print it out and use it as the basis for your designs, as shown in Figure 3.3.

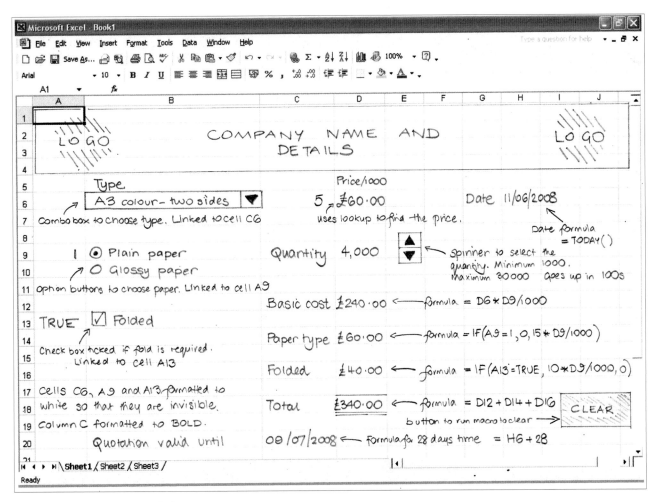

Figure 3.3 Quotation worksheet design

Produce a design for every worksheet you need to create. If you are using a formula, for example to add up a column of figures, indicate this on your design.

If you are using macro buttons, state what the macro will do when the button is selected. Annotate other features such as combo boxes, spinners or option buttons.

Indicate where any validation techniques will be used.

Consider printed output and security.

<tip>

When your designs are finished, you should be able to give them to a third party (such as someone else in your class) and they should be able to implement the solution if required.

Questions

◀

1. Produce a design for the Germany page of the Perfect Cottages website. Remember that there are 19 cottages. It needs to include the company logo, a thumbnail and a small amount of text for each cottage, and a link back to the home page.

2. Choose one of the scenarios below and produce the screen designs:

   - Uncle Horace wants to store his digital holiday photos in a format in which he can easily see them.
   - A local hot-food, take-away restaurant requires a website to display prices, photos, opening times, delivery costs and how to get to the restaurant.
   - The captain of Rottingdale Cricket Club wants a website to show visitors results of matches this season and forthcoming fixtures. They have two teams, the first XI and the second XI and they play matches every Saturday and Sunday throughout the summer.
   - The Bridge Hotel near Snowdon in North Wales would like a small website to advertise their 12 single and 23 double rooms.
   - The West Gower Youth Centre wants to set up a website so that youngsters can find out what is happening at the centre in the next month.

Selection and use of input devices and input media

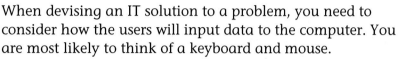

Input devices

When devising an IT solution to a problem, you need to consider how the users will input data to the computer. You are most likely to think of a keyboard and mouse.

However there are several other types of input devices that may be useful.

▶ Scanners

Scanners can be used to input an image. The most common form is a flatbed scanner that takes an image on paper that is usually up to A4 size. You can use a scanner to input a plan or a map, a chart or diagram, a photograph or even a signature.

Pictures can be stored in a number of formats such as JPG or BMP.

Scanners are often sold with optical character recognition (OCR) software (see page 31) that enables text to be scanned in and stored so that it can be loaded into a word-processing program. Some manufacturers claim at least 99 per cent accuracy but the accuracy of the output produced by software depends on the image quality of the original document.

Figure 4.1 Scanner

▶ Digital cameras

Digital cameras are widely used. The camera can connect to a computer via its USB port or a memory card can be removed from the camera and inserted into a card reader connected to the PC.

Images can be downloaded and stored. Image-processing software can be used to modify an image, perhaps by cropping, altering the brightness or contrast, or removing the "red eye" effect caused by the use of a flash. Photographs can then be printed out to a particular size, kept on storage media (such as CD-R) or displayed as a slide show on the screen.

Many digital cameras can record moving images and sound. Digital video cameras can also store video in a format that can be loaded into a PC.

Figure 4.2 Digital camera

Figure 4.3 Touchpad

▶ Touchpads

Laptops, which are often used in confined spaces where there isn't a flat surface for a mouse, may use a touchpad. A touchpad is a rectangular pad on a laptop keyboard located below the space bar. It is touch-sensitive and you move the cursor by moving your finger around the pad, as if you were drawing.

Some laptops use a rubberised nipple instead – it is a tiny rubber nub that sticks out of the keyboard between the G, H, and B keys. You move the cursor by moving the nipple like a joystick.

Neither input device is as easy to use as the standard mouse.

Figure 4.4 Graphics tablet

▶ Graphics tablets

A graphics tablet is a device with a special flat surface. You can draw on this surface with a stylus (similar to a pen) and the drawing is automatically read by the computer. These allow very accurate images to be produced and are ideal for graphic designers, artists and technical illustrators.

▶ Special keyboards

An **ergonomic keyboard** is an alternative keyboard that is reputed to reduce the risk of repetitive strain injury (RSI).

A **concept keyboard** is a keyboard designed for a particular purpose. Fast-food chains use such keyboards where there is a grid of images, one for each menu item. The operator merely has to press the image that represents the dish that has been chosen.

Figure 4.5 Ergonomic keyboard

▶ Touch screens

Touch screens allow users to make selections by apparently touching the screen. In fact there is a grid of infrared beams in front of the screen. Pointing at the screen breaks the beams so giving the position of the finger. A touch screen is both an input and an output device. They are commonly used in railway stations for the purchase of tickets, in museums and galleries for interactive use and in towns and cities to help the visitors find out more about the area. Visitors can select from a number of icons representing such things as places of interest, special events and transport. When an icon is touched, more information is displayed. A touch screen can provide a robust device that is very simple to use.

Typically, personal digital assistants (PDAs) use touch-screen and character-recognition technology to allow the user to choose from menus and "write" on the screen to enter text into the computer.

Figure 4.6 Touch screen

▶ Optical character recognition (OCR)

Optical character recognition is a method of data capture where the device recognises characters by light-sensing methods. These characters are usually typed or computer printed, but increasingly devices are able to recognise handwritten characters.

Traditionally, OCR has been used for many years to input large volumes of data by commercial organisations such as utility companies. OCR is now common using a flatbed scanner connected to a PC. Scanners are usually sold with OCR software as well. The system uses a two-stage process. First a page of text is scanned and the data is encoded as a graphical image. Then the OCR software scans the image using special pattern-recognition software where groups of dots are matched against stored templates. A character is selected when a sufficiently close match is found.

Figure 4.7 Close-up shot of an optical character reader used by the Post Office as part of its automatic mail sorting system

case study 1	A version of OCR called automatic number-plate recognition (ANPR) is used by the police to read car-registration numbers picked up by CCTV or speed cameras.
▶ OCR on the roads	As a vehicle approaches the camera the software takes a series of "snapshots" and stores them in a file. When the number plate is of sufficient size for the OCR software, the frame is scanned and the registration number is read and stored.

A similar system is used in London in the congestion charge zone. Motorists entering this zone have to pay a charge, which was initially £5 per day. The system cost £200 million to implement but the scheme paid for itself within 30 months. It has 230 cameras on eight-metre poles. The cameras capture the registration numbers of 98% of the cars in the charging zone.

The police use ANPR to check for stolen cars and to catch speeding motorists. As a vehicle passes, the ANPR equipment reads the number plate and the software checks it against the Police National Computer database. If the number plate is matched, for example, with a stolen car, the ANPR equipment sounds an alert so that the car can be stopped.

Figure 4.8 OMR form

▶ Optical mark recognition (OMR)

Optical mark recognition is a form of mark sensing where pre-printed documents are used. These documents contain boxes where marks can be made to indicate choices, or a series of letters or numbers that can be selected to indicate choices. The reading device detects the written marks on the page by shining a light and recording the amount of light detected from different parts of the form. The dark marks reflect less light. The OMR reader transmits the data about each of the pre-programmed areas of the form where marks can be expected. The forms are printed with a light ink which is not detected.

Multiple-choice exam papers use a special OMR answer sheet. Each student is given a mark sheet and they put a mark with a pencil through their choice for each answer (usually represented by a letter) on a grid. All answer sheets are gathered together and read by an OMR reader. Appropriate software checks the marks against the expected ones for the correct answers and the students' scores can be calculated.

The National Lottery and football pools coupons use a similar system. In some hospitals, patients can select meals from a menu printed on OMR cards. The cards are fed into a computer system that produces accurate information quickly to be used by the catering services. OMR is a popular method for collecting questionnaire data.

OMR avoids human keyboard entry, which is prone to mistakes. OMR allows for fast entry of a high volume of data.

Forms have to be carefully designed and filled in as marks that are entered in the wrong place may be ignored or may cause the form to be rejected. Care needs to be taken to keep the forms uncreased, as bent or damaged cards are likely to be rejected by the reader.

Figure 4.9 MICR document

► Magnetic ink character recognition (MICR)

Magnetic ink character recognition is another fast and reliable method of entering data. Documents are pre-printed with character data in special ink that can be magnetised. The print appears very black. The shape of the characters is recognised by detecting the magnetisation of the ink. MICR is almost exclusively used to read cheques. The bank sort code and customer account number are printed in magnetic ink. The banks' clearing house computers can read these numbers. It is used instead of OCR to help minimise fraud. The hardware is expensive, so MICR is only used in situations when security is important.

Figure 4.10 Bar code and reader

► Bar-code readers (wands and scanners)

Bar codes are used as a means of identification of items in a store, every product having a unique article number. The pattern of the black and white stripes represents a code. The codes are read from the stripes by detecting patterns of reflected light and encoded into computer-readable form using some form of scanner. Each number in the bar code is represented by four stripes (two black and two white). The bar code has separate right and left sides and can be read in either direction.

Bar-code readers can be hand held (wand), where the reader is moved over the bar code, or fixed (scanner) where the object is moved across the reader. Laser scanners are frequently used in supermarkets to read the bar code which identifies the product.

The bar code can be read very quickly with very few mistakes. A number is often printed beneath the bar-coded representation of the code. If, for some reason, the bar-code reader fails to read the bar code, the operator can key in the code for the product.

The data from the bar codes is then used to calculate the shopper's bill and keep a record of items sold and stock levels.

Two common numbering systems are the European article number (EAN) and the universal product code (UPC). These numbers are structured so that each part of the code has a particular meaning. For example, an EAN product code consists of 13 digits. The first two stand for the country of origin, the next five are the manufacturer's number and a further six identify a specific product. (The last of these six is a check digit.)

Figure 4.11 Magnetic-stripe card

▶ Magnetic-stripe card

A magnetic-stripe card has a strip of magnetic material on the back of a card, such as a credit, debit, cash or loyalty card, that is usually made of plastic. Encoded information stored on the magnetic stripe on the back of the cards can be automatically read into a computer when the card is "swiped" through a reader. The pattern of magnetisation is detected and converted into encoded binary form for computer use. This process is much quicker than typing in the card number and is accurate, not being susceptible to human error. The stripe can be damaged if the card is exposed to a strong magnet as the magnetic pattern would be altered and so render the card unreadable.

Magnetic-stripe cards are widely used on train tickets. The stripe holds data concerning the journey. On the London Underground, tickets are read when a passenger arrives at, or leaves, a station. As well as being used to confirm the validity of the ticket, data is collected on all the journeys made. This can be used to produce information on passenger patterns and flows that can be used when planning services.

When parking a car at Heathrow airport, drivers are issued with a ticket that has a magnetic stripe on the reverse. At the time of issue, the date and time of arrival are encoded onto the strip. When leaving the airport building to return to the car park, the driver inserts the ticket into a large payment machine. The machine has a magnetic-stripe reader that reads the data from the ticket. The charge, based on the length of stay, is calculated and displayed on a small screen. After the driver has made the payment, using cash, credit or debit card, details of the payment are recorded onto the strip on the card, which is then returned to the driver. When the driver leaves the car park in his car, he inserts the updated ticket into a machine controlling the barrier.

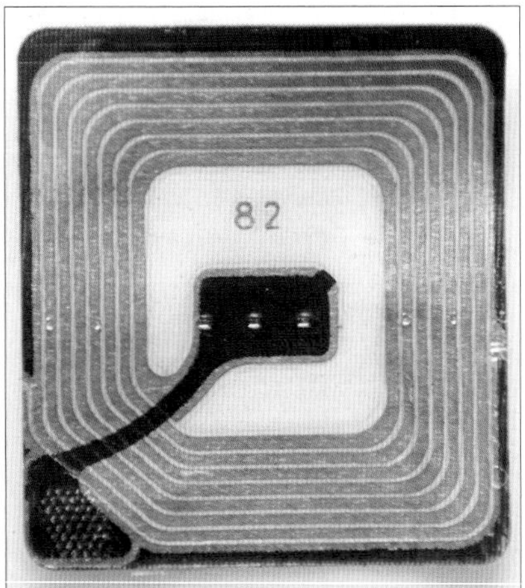

▶ Radio-frequency identification (RFID)

RFID involves tags that emit radio signals. Different tags emit different signals that can be recognised by readers. It is not necessary for the tag to touch the reader, as tags can be read from around 65 cm, and the reader does not need line of sight.

As a result, RFID is a fast and reliable method of identification. It is used in the following applications:

- New UK passports can immediately be checked against a list of stolen passports, suspected terrorists, etc.
- In London, bus and underground passengers can use an Oyster card to pay for journeys. The passenger just has to put the Oyster card on to a yellow reader at the station or on the bus. Find out more at http://www.oystercard.com.

Figure 4.12 RFID tag

▼

- Distribution companies can track deliveries using high-frequency RFID tags. They can provide customers with real-time information about the progress of a delivery.
- Car manufacturers supply RFID-equipped ignition keys as an anti-theft measure. The driver can unlock the doors and start the car without taking the key out of their bag or pocket.
- RFID tags can be implanted and used for animal identification.

▶ Touch-tone telephones

A method which is being increasingly used for limited data capture is a touch-tone telephone. Banking systems allow the user to enter the account number using the keys on the telephone handset, in response to voice commands. The user can also choose the type of transaction required by pressing a specific key.

example ▶ dialogue with an automated system	*Computer system:* Thank you for calling the bank. Please enter your account number followed by #. *User enters 21978433#.* *Computer system:* Thank you. Please enter your sort code followed by #. *User enters 189345#.* *Computer system:* Thank you. Please enter your four-digit security code followed by #. *User enters 2989#.*

▶ Speech recognition

Speech recognition is a growing area of computer input. As computers get faster and memory increases, reliable speech recognition has become cheaper. Already mobile phones can recognise the name of the person the user wants to call. Soon we will be able to use speech technology to access information without the need for a keyboard or a mouse.

There are two main uses made of speech recognition systems. They are used to enable large quantities of text data to be input as words into a computer so that they can be used in a software package such as a word processor. They can be used to input simple control instructions to manipulate data or control software. This is particularly appropriate when the environment makes the use of a keyboard or mouse unsuitable, such as in a factory. Users of such systems can create and modify documents in a word processor. They can enter data into a variety of packages including form filling. The user can surf the Web or send emails simply by speaking.

Speech recognition systems provide users who have a limited ability to type, possibly due to repetitive strain injury or other disabilities, with a manageable way of interacting with a computer system.

Speech recognition systems are used to enter large volumes of text and are particularly useful for professionals who are not skilled in keyboard data entry.

In the past such users would have dictated to a secretary or used a Dictaphone so that an audio-typist could later type the text. The use of speech recognition software allows the user to input the text directly without the need for a typist. This removes any time delay and provides the user with control.

Advantages of using a speech recognition system

- Users who are not trained typists can achieve faster data entry than through keyboard use.
- The user's hands are free for other purposes, such as driving a car.
- Users with disabilities can enter data.
- It can be used in areas where a keyboard is impractical.

Limitations of speech recognition in practice

- The user must speak each word separately leaving clear gaps between the words. This is not how they would speak naturally.
- The software needs to be trained to recognise an individual voice.
- Excessive background noise can interfere with interpretation of speech.
- Homophones (words which sound the same but are spelt differently, such as their, they're or there) can be confused.
- Many cheaper systems only recognise a small number of words.

---

case study 2
▶ **Dragon Naturally Speaking**

Dragon NaturallySpeaking claims to be the best-selling speech-recognition software in the world.

It claims to work at speeds of up to 160 words per minute – about three times faster than typing – with accuracy rates as high as 99%.

It allows the user to dictate directly into popular software such as Microsoft Word, Microsoft Outlook Express, Microsoft Excel and Corel WordPerfect.

Activity 1

Find out more at http://www.nuance.com/naturallyspeaking/.
What does the website say are the advantages of using NaturallySpeaking?

Which input device and input medium should I use? ◄

The input device chosen is influenced by factors such as cost, speed and accuracy.

► Volume of data

If the volume of data is large, then automatic equipment is appropriate. The cost of installation, maintenance and staff training can be justified. With a high volume of data, manual data entry such as a keyboard would not be able to cope within a reasonable time.

► Speed

Speed can be an important factor in data entry. A fixed bar-code scanner can be used at higher speeds at a busy supermarket checkout than a hand-held scanner. Automated data entry methods, such as OCR and OMR, can produce very fast data entry.

► Nature of system

A particular system may have a specialised input method associated with it. For example, the need for security against fraud means that magnetic ink character recognition (MICR) is the chosen method for entering data from cheques.

In a dirty or dangerous environment, such as a factory, the use of a keyboard may not be appropriate. Stock and goods may be identified by bar code. The use of magnetic media may be impractical due to magnetic fields.

► Ease of use

The conditions under which data is to be entered, together with the range of skills of the users, may influence the choice of data capture method.

► Technological development

The choice of data entry method for a particular system may have been different a few years ago and may be different again in a few years' time. New methods are being developed all the time. However, it is important that methods are reliable and have the confidence of the users.

► Cost

The cost of installing a particular input method is a major factor in any decision. Cost could relate to staffing or hardware, as well as the cost of changing from the old system to the new. The greater the volume of data, the more likely that a higher cost method can be justified.

Activity 2

Match the input devices and uses listed below and explain your choices:

1.	Inputting a photograph	**a.**	Bar-code reader
2.	Reading the details from a bank cheque	**b.**	Scanner
		c.	MICR
3.	Reading the turnaround document from the bottom of a gas bill	**d.**	OMR
		e.	OCR
4.	Identifying a tin of peas		
5.	Inputting the answers to a questionnaire		

Here's another set:

1.	Playing a game	**a.**	Graphics tablet
2.	Finding out information about bus times at a bus station	**b.**	Concept keyboard
		c.	Touchpad
3.	Entering the design for a garden	**d.**	Joystick
4.	Selecting a vegetable type at a supermarket checkout	**e.**	Touch screen
5.	Cutting and pasting text using a word processor on a laptop		

Activity 3

On the Internet, perform a search on OCR scanner accuracy rate. Can you find out how accurate OCR scanners are?

Activity 4

Visit this site for reviews of different input devices: http://reviews.cnet.com/

Automatic input methods, such as scanning a bar code, remove the need to key in data. They allow large quantities of data to be read quickly and accurately.

A range of input methods have been developed, each suited to a different range of applications and circumstances:

▶ Direct entry using a keyboard

▶ OCR

▶ OMR

▶ MICR

▶ RFID

▶ bar-code scanning

▶ magnetic-stripe card

▶ touch-tone telephone

▶ speech recognition.

There are a number of factors that must be considered when choosing an input method for a particular system. These include:

▶ the volume of the data

▶ the speed at which it has to be input

▶ the nature of the system

▶ the ease of use

▶ the state of current technological development

▶ the cost.

Questions

◀

1. Name a method of data capture, other than the use of a keyboard, that would be suitable for capturing the following data:
   a) marks from multiple choice examination answer sheets (1)
   b) details from bank cheques (1)
   c) details of library book loans and returns. (1)
   June 2007 ICT2

2. Give two different methods of entering data into a computer system. (2)
   June 2006 ICT2

3. Describe two advantages of using a scanner and OCR software rather than a keyboard to enter text. (4)

4. A village store has just installed an electronic point of sale (EPOS) system including a bar-code reader.
   a) Describe two advantages that the store gains from using the EPOS system. (4)
   b) Describe one disadvantage to the store of the EPOS system. (2)
   January 2005 ICT2

5. Pictures are scanned onto a computer and stored in files. Describe two types of files that can be used to store pictures on a computer. (4)

6. A friend has told you that she has bought a scanner with OCR software and software that can convert photos to JPEG. However, she does not know what these terms mean.
   a) What does the term OCR mean? (1)
   b) What does the term JPEG mean? (1)

7. A pizza delivery man uses a satellite navigation device. State one input device suitable for a satellite navigation system. (1)
   Explain why your chosen input device is more suitable than
   a) a QWERTY keyboard (2)
   b) a mouse. (2)

8. The local council wishes to store the contents of documents on a computer system. The documents consist of handwritten and typed text. The documents will be scanned and OCR software will be used to interpret the text and export it to files that can be read by word-processing software.
   a) Describe two problems that could occur when scanning and interpreting the text. (4)
   b) Describe two advantages to be gained by using OCR software. (4)
   June 2001 ICT2

9. Automatic Teller Machines (ATMs) are the cash machines that we use outside banks and supermarkets to withdraw cash or to check our bank balance. As these devices are outside, they need to be weather-proof and secure. They use the following input devices:
   ■ A magnetic-stripe reader
   ■ A specialised keypad that allows numbers to be entered
   ■ Keys that are used to make selections from a menu.
   Describe an example of what each input device is used for. (6)

Selection and use of storage requirements, media and devices

AQA Unit 1 Section 5

Why data is stored

Storage of data is an essential part of ICT. Data is likely to be used again in the future and entering it every time it is needed would be very time-consuming.

Storage media may be used when purchasing software or when transferring data from one computer to another. Data is also stored as a backup in case the original stored data is damaged, for example, by hardware failure or a virus.

Data is also archived. This means removing data that is not used frequently from the main backing storage device of the computer (usually a hard disk) to a medium that can be stored away from the computer, but accessed if the data is needed. At a college, the details of students who left college three years ago would not be required on a day-to-day basis. However, they may need to be accessed occasionally, for example, if a request for a reference about a student was received.

A home computer user who is a keen photographer could soon fill up the hard disk with digital images. To free up space, without deleting the images and losing them permanently, the user should archive some of the images to a removable storage medium.

Common storage devices

These devices all store data for later use. When the computer is switched off, data stored in main memory is lost but data stored in these devices is not lost.

▶ **Magnetic storage devices**

Hard disk

The most common storage device is a hard disk, sometimes called a hard drive. This is a magnetic storage device that is generally housed within the box containing the processor (an internal hard disk). The internal hard disk can hold many gigabytes of data. It is used to store the operating system and other software as well as data files.

A hard disk can be divided into folders to make it easier to store and find work. In Figure 5.1, a student has set up a different folder for each of the subjects that they are studying. Folders can be divided into sub-folders. In Figure 5.1, the student has sub-divided their ICT folder into a folder for Unit 1 and a folder for Unit 2.

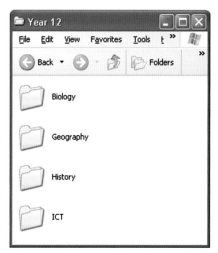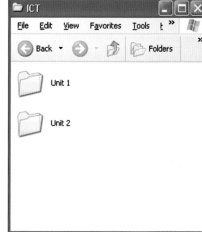

Figure 5.1 A hard disk divided into folders and sub-folders

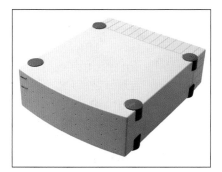

Figure 5.2 External hard disk

Figure 5.3 Floppy disk

External hard disk

It is increasingly common for a computer system to have an external hard disk which connects to the computer via a USB port. This hard disk can be used as a backup in case of internal hard disk failure. It can also be used to increase the storage capacity of a computer, if the internal hard disk is getting full. As it is small and relatively portable, it can be disconnected and connected to a different computer. This allows it to be used to transfer data from one computer to another.

Floppy disk

The removable 3½-inch floppy disk is a magnetic storage medium consisting of a small, flexible plastic, magnetic disk inside a hard plastic case. Their usefulness has decreased as the size of files to be stored has grown and the capacity of the disk is very small (normally 1.44 MB). Data transfer is also slow.

For many years, a floppy disk drive was an essential part of all personal computer systems. It was mainly used for keeping backup copies of small files, such as text files of letters or small documents, or for transferring small files between two or more computers. Floppy disks are still very cheap to buy but flash memory has become a more convenient way of storing files or taking them to another PC.

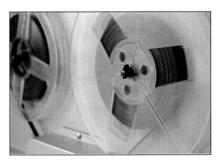

Figure 5.4 Magnetic tape drive

Figure 5.5 CD- and DVD-ROMs

Magnetic tape

Magnetic tape cartridges are used for backing up entire networks due to their high capacity, for example, 120 GB. Magnetic tape is a serial medium. This means that the data has to be read in the order in which it is stored on the tape. It is also sometimes used for archiving large volumes of data.

▶ Optical storage devices

CD-ROMs (Compact Disk–Read-Only Memory) are small plastic optical disks. Like floppy disks, CD-ROMs can be moved from computer to computer. However, they store larger quantities of data (650 MB or 0.65 GB) permanently. The data stored on a CD-ROM is less prone to damage than that on a floppy disk as the data-storage method uses laser rather than magnetic technology.

As the name suggests, data can be read from but not written to a CD-ROM disk. Much commercial software is distributed on CD-ROM and many computing magazines are sold with a CD-ROM that holds free sample software.

CD-Rs (Compact Disk–Recordable) are writable compact disks. Using a special writable CD drive, up to 650 MB of data can be recorded onto a blank disk. This data cannot be altered but it can be read by any other PC with a standard CD-ROM drive. This means that CD-R is suitable for archiving data that can be removed from the computer's hard drive, but is still available if it is needed.

CD-Rs are useful for storing multimedia applications as the sound, graphics and animation files are likely to require large amounts of storage space. For example, a marketing manager might prepare a presentation with information about new products and pricing policy which could be distributed to all sales staff.

CD-RWs (Compact Disk–Rewritable) are rewritable compact disks. Using the same writable CD drive, up to 650 MB of data can be recorded, deleted and re-recorded on these disks. This means that a CD-RW can be used in the same way as a floppy disk without the limitation on size of file.

DVD (Digital Versatile Disk) is a high-capacity optical disk developed in the 1990s. A DVD is the same physical size as a CD. Current disks can store 4.7 to 17 GB, the equivalent of over 25 CD-ROMs. They are used for films as well as audio.

DVD drives can read both CDs and DVDs. Combination (Combi) drives that read DVDs but write to CD-R and CD-RW are increasingly common. Some DVD drives can both read and write DVDs.

DVD-R (Recordable DVD) and **DVD-RW (Rewritable DVD)** are available but capacity is reduced.

▶ Flash memory

Flash memory is a type of removable backing storage that was developed for small devices such as digital cameras and is now used extensively as a portable storage medium.

There are a number of proprietary formats which require special adaptors e.g. Compact Flash and Secure Digital. Flash memory is made up of a special storage microchip. Flash memory sizes range from 8 MB to 4 GB.

A **USB memory stick** or pen drive is a portable storage device that uses flash memory. It is "plug and play": the user simply plugs the stick into their computer's USB port and the computer's operating system recognises the stick as a removable drive. A memory stick is very small and light; it can be attached to a key ring or hung on a cord around the user's neck. Memory sticks do not require any battery or cables and are available in capacities ranging from 128 MB to 4 GB. However, memory sticks with greater capacities will soon be developed.

Memory sticks are used to record various types of data: graphical images, music and moving pictures as well as other computer data files. A 1 GB memory stick can store between 250 and 1000 minutes of video depending on the image resolution chosen. Many MP3 players can also be used as memory sticks.

Figure 5.6 USB memory stick

Activity 1 – storing your work

You will need to store your work for this and other courses. You may need to load it onto your home computer. You may wish to keep a backup if it is very important and losing it could be detrimental to your studies.

1. Create the grid with the headings given below.

Device/medium	Description	Applications
Floppy disk		
Memory stick		
Hard drive		
CD-RW		

2. In the description column, describe the storage medium.

3. In the applications column, describe when you might use this medium as part of this course.

Activity 2 – which medium?

For each of the following activities, state a suitable backing storage medium. Justify your choice in each case.

- A software house distributing new software
- A businessman transferring a short text document from his laptop to his desktop computer
- Storing music downloaded (legally) via the Internet
- Storing frequently used software
- A charity archiving details of past transactions from its database
- An A-level student transferring an ICT project between home and school computers
- Backing up data from a laptop
- Backing up all the data stored on a network
- Storing holiday photographs
- Storing a multimedia presentation about your sixth form for distribution to local schools.

Activity 3 – capacity

Visit the websites of manufacturers or suppliers of computer hardware. Investigate the current maximum capacity for a memory stick, a magnetic tape and a hard disk.

▼

SUMMARY

Storage devices can be fixed or removable:

▶ Hard disk provides very high capacity (10 to 120 GB), fixed storage. It allows permanent online storage of software and data.

▶ Removable backing storage is used for increasing capacity, backup, archiving, and transferring software and data between computers.

▶ The floppy disk has a low capacity of 1.44 MB and is used to transfer small files between computers.

▶ A magnetic tape cartridge has a very high capacity (e.g. 120 GB). It can be used to back up an entire network or for archiving large volumes of data.

▶ Flash media have a medium capacity (128 MB to 4 GB) and are used for the transfer and backup of large files.

▶ A CD-ROM has a high capacity (650 MB). It is read only and is used for distributing software and reference material.

▶ CD-R and CD-RW have a high capacity and can be used to archive, transfer and back up all types of data. They are commonly used for storing multimedia presentations for distribution.

▶ A DVD is read only. It has a very high capacity (4.7 to 17 GB). It is mainly used to distribute moving image and audio data.

▶ A DVD-R has a high capacity (4.7 GB) and can be used to record moving image and audio data.

Questions

1. Choose a suitable storage medium from the list below for:

   a) Storing your work.   (1)

      Give a reason.   (1)

   b) Storing a podcast.   (1)

      Give a reason.   (1)

   c) Keeping a blog.   (1)

      Give a reason.   (1)

DVD-ROM, Flash Memory, CD-ROM, CD-RW, Portable Media Player, 500 GB External Hard Drive

AQA Specimen Paper 1

2. It is necessary to back up files that are stored on a computer system. For each of the following files, state a medium that may be used for backing up. Explain why it is the most appropriate medium.

   a) A multimedia presentation   (2)

   b) A word-processed letter   (2)

   c) A student's ICT coursework that consists of text and graphical images.   (2)

3. Floppy disks were once the usual medium for distributing software adopted by software houses and developers.

   a) State three reasons why the use of floppy disks has decreased.   (3)

   b) Describe one occasion where it would still be sensible to use a floppy disk for software distribution.   (2)

   c) State two ways, other than by floppy disk, in which software can be distributed. Give an example, with reasons, of when each one might be used.   (6)

4. Modern personal computer systems usually include a CD rewriter. State two legal uses for a CD rewriter.   (2)

June 2002 ICT2

5. You have to design, create and test solutions to solve clients' problems. To complete this project you should use a number of storage media.

   a) Suggest three different storage media which would be useful in creating the solution.   (3)

   b) For each of your answers to part (a), describe a purpose for which you would use the different storage media.   (6)

6. A student told a teacher, "I'm sorry but I can't hand in my project work as it was all on computer. My computer at home has crashed and my files are lost." Suggest a response from the teacher.   (3)

7. Sally is an author who has recently purchased an external hard disk drive that plugs into her laptop computer using the USB port. She is now producing some chapters for a new book.

   a) Explain one reason for saving her files to her internal hard disk rather than the external hard drive.   (2)

   b) Explain one reason for saving her files to her external hard disk rather than the internal hard drive.   (2)

8. Computer users often store datafiles in folders and subfolders. Jack is the secretary of the local Model Aeroplane Club and often has to write letters and reports on behalf of the club.

Jack's wife, Helen, is the treasurer of the local brownie pack. She uses the same computer to produce accounts and balance sheets. Draw a suitable file structure for Jack and Helen to store their data.   (4)

Selection and use of output methods, media and devices

AQA Unit 1 Section 6

When devising your solutions to the problems you stated in Chapter 2, you need to consider how the user will get information from the computer. The most familiar methods would be to use a monitor (screen) or a printer.

However, there are different types of monitor and printer and several other types of output devices, such as plotters and loudspeakers, that you should consider.

Screens

The most commonly used output device is the screen. Screens can be black and white or colour and vary in size, resolution and type. The screen size is measured from one corner to the diagonally opposite corner. Common screen sizes for desktop computer screens are 12, 14, 17, 19, and 21 inches.

Smaller screens can be used on portable devices such as PDAs or mobile phones. Where larger screens are required, computers can be connected to flat-screen, liquid crystal displays (LCD) or plasma televisions, which work in a similar way. 42-inch screens are not uncommon and even bigger screens exist.

The screen's resolution determines the number of dots (known as pixels) that are displayed – the more pixels, the greater the resolution and the better the picture quality.

Traditional screens make use of cathode ray tube (CRT) technology and are bulky devices. Flat screens were originally used in laptop computers because of their light weight and compact size. Now they have become very common with desktop computers. Their price has reduced considerably and their quality has increased. They take up much less desk space than a CRT monitor, they use less electricity and generate less heat – this can be very important in a stuffy office.

Modern flat screens have bright displays and high resolutions. They are a type of LCD screen that uses a technology known as thin-film transistor (TFT).

A problem with LCD displays is that defects can occasionally occur in the manufacturing process. The screen is made up of a number of pixels. Each pixel is made from three sub-pixels: one red, one blue and one green. Pixel defects can occur at any stage in the LCD's life and cannot be fixed or repaired.

Figure 6.1 CRT and LCD monitors

Printers

When choosing a printer, thought needs to be given to what it will be used for as this will determine the most appropriate type of printer. The volume of printing required will influence the choice as the speed of printers can vary enormously. A home user who prints out the occasional letter and colour photograph has very different needs from a large organisation, such as a bank that produces hundreds of thousands of statements in a day.

Some applications require the use of pre-printed stationery such as headed notepaper or pay slips which are directly printed into envelopes so that the contents of the slip cannot be read casually by anyone. Sometimes multipart paper is used; invoices for car repairs at a garage are often printed on two-part paper with one copy for the customer and one for the garage.

Printers come in different sizes. The printer at your school or college may well be larger than a printer you would buy for home use as the volume of printing done at school is likely to be much larger and so a more powerful printer is needed. However, much smaller and much larger printers are in use. Very small printers can be built into equipment for specialist use. Lightweight, portable printers can be used with laptop computers. At the other end of the scale, high-speed laser printers are floor-standing and very hefty pieces of equipment. They are often used for continual printing.

The most common forms of printers all form their images out of dots – the smaller the dots, the better the quality of the print.

Figure 6.2 Dot matrix printer

▶ Dot matrix printers

Dot matrix printers are a type of impact or contact printer. They print by hammering pins against a ribbon on to the paper to print the dots. They are most commonly used in situations when only text, in black and white, is required. As it is an impact printer, a dot matrix printer can be used to print on multipart stationery; the pressure of the pins cause the carbon paper to produce a print. These printers are easy and cheap to maintain as replacement ribbons are inexpensive. However they can be noisy and the quality of print is generally not as good as with inkjet or laser printers.

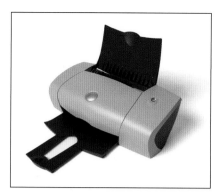

Figure 6.3 Inkjet printer

Figure 6.4 Laser printer

▶ Inkjet printers

Inkjet printers squirt ink on to the paper and form letters from tiny dots. They are quiet, quality is good and most versions support colour printing. Such printers are the most popular for home use as they are relatively cheap to purchase and produce a clear image. However, most inkjet printers are too slow to be used with high volumes of printing. On some inkjet printers, photo quality images can be printed on special high quality, glossy paper. Although the initial cost outlay is relatively low, the ongoing cost of cartridges can be very costly in relation to the initial cost of the printer.

▶ Laser printers

Laser printers are very popular for business use. The way in which a laser printer functions is very similar to the way a photocopier works. In fact, some printers can also function as photocopiers.

The cost of using a laser printer depends on a combination of costs: paper, toner replacement and drum replacement. Many laser printer models can accommodate a duplexing unit that allows printing on both sides of the paper in one run. This reduces paper costs and the space required for physical filing of documents.

The variety of types of laser printer available is huge. Printers that can produce colour printing are more expensive than those that can only produce black and white printing. Speeds can vary. The slowest can sit on a desk top and are used occasionally to produce letters and brief documents at a speed from four pages a minute. The fastest are large, high-speed devices that can print hundreds of pages in a minute and are used, for example, by banks to print out statements.

The choice between an inkjet and a laser printer is not always clear cut. Many inkjet printers can produce excellent quality printouts. Printer choice can be a compromise between print quality, speed and cost. Some of the more expensive inkjet printers produce better quality graphics than some laser printers.

Activity I

Use the Internet to investigate the similarities and differences between laser and inkjet printers. In particular, investigate the price and speed of different types and makes of printers.

► Thermal transfer printers

A thermal transfer printer is a non-impact printer that uses heat to make an impression on paper. They operate quietly. Thermal transfer printers are particularly used for printing bar codes, price tags and labels. Thermal transfer printers are usually small and are cheap to maintain.

A portable version of a thermal printer can be used for mobile point of sale (POS), receipt or ticket printing. The printer is very small – only 58 mm wide – and runs off a battery. These printers can be found in restaurants, used when customers make a payment by credit or debit card. The printer is used to produce a receipt at the table.

► Dye-sublimation printers

Dye-sublimation printers (called dye-subs) are used to produce high-quality graphical images, particularly colour photographs. These printers produce images with excellent colour reproduction, that look as if they came from a photographic laboratory. The purchase price of a dye-sublimation printer is high; this is a printer that would be used for printing artwork to a very high quality.

Data projectors ◄

Figure 6.5 Interactive whiteboard in a classroom

Data projectors are expensive but are now common in business demonstrations, where they usually project the output from a laptop computer onto a large screen. Prices continue to fall and quality is improving so that they are bright enough to be used in a room lit by natural light.

They are increasingly used in the classroom, often in conjunction with an electronic interactive whiteboard.

Speakers

◀

Figure 6.6 Speakers

Speakers are a common device used to output music. However, sound is used in other ways as well. Many programs make audible warnings – beeping sounds to indicate errors or as an alert, for example when you press the CAPS LOCK key. Microsoft Windows makes a distinctive sound when loading, by default.

The use of a speaker for speech synthesis is an expanding area of output, common in a range of applications including computer games. If you phone directory enquiries, you are told the number you require in synthesised speech. Have you ever wondered who calls out "Cashier number four, please" in shops, banks and post offices?

Headsets

◀

Figure 6.7 Headset

Headsets combine an input device (a microphone) and an output device (loudspeakers). With the development of VoIP (Voice over Internet Protocol), millions of computer users can make free telephone calls to other users over the Internet often using a headset.

Activity 2 – garden design

Terry has built up a small business as a garden designer. He wishes to buy a computer to help him to run his business. He intends to produce letters, keep records on his customers and jobs, create garden designs and create leaflets showing "before" and "after" shots of the gardens he produces. Some of these photographs are taken by Terry and some are given to him by his clients. Terry wishes to store all the photographs and designs electronically.

List the input, output and backing storage devices that Terry should buy, explaining carefully why each is needed.

case study 1
▶ at the supermarket till

Figure 6.8 EPOS at a supermarket

The checkout at the supermarket has a large number of output devices:

■ As products are scanned, a screen displays details of the latest purchase together with the current total cost for the customer to view.

■ A speaker makes a beeping sound to indicate that the item has been scanned correctly.

■ A small LCD display on the chip-and-PIN device gives instructions to the customer.

■ A small, built-in printer is used to print out the bill which lists every item purchased together with its price.

Can you think of any other output devices they use?

SUMMARY

A variety of output devices are used to present information:

▶ **printers of various different types**

▶ **monitors of various types and sizes**

▶ **data projectors**

▶ **speakers.**

The choice will depend on the use that is made of the information.

Questions

◀

I. A hotel is purchasing new computer hardware and software.

    a) State two types of printer that the company could purchase. (2)

    b) For each type of printer, describe an appropriate use within the hotel. (4)

    c) Name two input devices that would be used. (2)

    d) The hotel manager is considering purchasing himself a personal digital assistant (PDA). Describe one output device and two input devices he would expect to find on his PDA. (6)

2. Many cars today use satellite navigation devices.

    a) Suggest two possible output devices for a satellite navigation system. (2)

    b) Explain why your answers to part (a) are more suitable than a standard computer monitor and a laser printer. (4)

3. A company has offices on five sites. Each office has between 10 and 20 members of staff working in it. Internal email is used as a means of communicating between the staff. It has been suggested that speech recognition input and voice output might be used for the email system.

    a) State the extra input and output devices each PC would need to support speech recognition input and voice output. (2)

    b) State two advantages to the staff of using a speech recognition system. (2)

    c) State three reasons why the speech recognition system may not be effective. (3)

    d) State two disadvantages of the voice output system. (2)

January 2002 ICT2

4. Jake notices that someone locally is selling a brand new dot-matrix printer. The price of this printer seems very cheap. Jake's local chain store sells only inkjet printers and cartridges which are more expensive than the dot-matrix printer. Advise Jake on what he should purchase explaining two reasons for your choice. (4)

5. Restaurant owners use a wireless, hand-held computer to receive payment from customers. This computer is taken to the table, the customer inserts a credit card and enters their PIN. If the PIN is correct, a message of confirmation is displayed on a small LCD screen and a built-in thermal printer is used to print the customer's receipt.

    a) Describe two input devices. (4)

    b) Describe two output devices. (4)

6. State three output devices you might find in an automatic teller machine (ATM). (3)

7. An MP3 player usually has two output devices. What are they? (2)

Selection and use of appropriate software

AQA Unit 1 Section 7

Software

When devising your solution to the problem you chose in Chapter 2, you need to consider which software package is the most appropriate.

It may be appropriate to use more than one package.

▶ What is software?

Software is the name given to computer programs. Programs are made up of thousands of instructions that actually make the computer do what is required. Without software, computer hardware is no use. Programs are needed to control the hardware, the physical components of the computer, such as a printer, a processor or a disk drive.

▶ Types of software

There are two main types of software and each can be divided into subcategories:

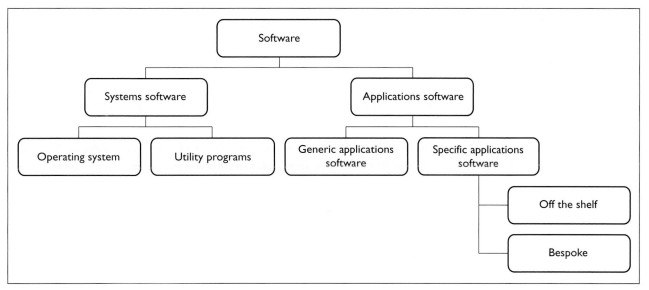

Figure 7.1 Types of software

■ **Systems software** is the name given to the programs which help the user to control and make the best use of the computer hardware. The operating system and utility programs are types of systems software.

- **Applications software** is a term used to describe programs that have been written to help the user carry out a specific task, such as paying wages, designing a newsletter or storing details of payments. This task is usually something that would need to be done even if a computer was not used. A spreadsheet package and a booking system program written for a dental surgery are examples of applications software.

Systems software ◀

▶ Operating systems

An operating system is software that is needed to run a computer. It controls and supervises the computer's hardware and supports all other software. It hides the complexity of the hardware from the user and provides an interface between the computer's hardware and the user and applications. This means that the user does not have to bother about the details of the hardware. An operating system manages the computer's resources: memory, storage, processor time, files and users.

▶ Utility programs

Another category of systems software is utility programs. These are also known as housekeeping programs as they allow the user to carry out tasks that make the use of the computer easier such as tidying up the storage of files on a hard disk.

Utility programs are tools to help the user make more effective use of a computer. Although utility programs are not part of an operating system, they are often provided with operating system software. Examples of utilities include:

- **virus-checking** programs scan a computer's storage devices for the presence of known viruses.
- **file-conversion** programs convert files from one format to another. For example, a document file produced by Microsoft Word is stored with formatting codes that are specific to Word. For the document to be read by WordPerfect, a different word-processing package with different formatting codes, a file-conversion utility must be used.
- **disk-formatting** programs set up a new disk so that the computer's operating system can access it. Disk formatting was a common occurrence for users when floppy disks were the main form of transferable backing storage and it is still sometimes necessary for rewritable CDs and DVDs.

- **file-compression** (sometimes referred to as zipping) programs reduce the size of a file. This is done to reduce the amount of backing storage space needed. Smaller files are quicker to send over the Internet.
- **file-management** programs enable the user to delete, rename or move files held on a backing storage device from one folder to another; copies may be made within different folders on the same device or on a different medium for backup or transfer purposes; new folders (directories) can be created.
- **backup** programs make a copy of a file for security purposes.
- **garbage-collection** and **defragmentation** programs remove unwanted data and files from a disk and close up any gaps left on the disk so that all the free space is together. This makes more room to store new files and enables the computer to operate more efficiently.

Applications software ◀

Applications software is software that has been written for a user task, such as producing invoices, cropping a photographic image or booking a flight.

Applications software can be package (or 'off the shelf') software, which is software that is developed and sold to a number of users. Such software is usually supported with a range of manuals or guides. This type of software can be bought in a shop, from websites, through trade magazines or direct from the supplier. Package software falls into two categories: generic and specific.

▶ Generic applications software

The most common form of software is generic or general-purpose applications software. Generic applications software is an application package that is appropriate to many users. Many such programs are pre-installed on a computer when it is sold and can be used in many ways. You will use this type of software to implement the solutions to the problems you chose in Chapter 2.

Many generic software packages can be used to create customised applications for a user. The use of macros, buttons and customised toolbars enable this to be done. Data can often be imported from one type of package to another.

Word processing, database management, spreadsheet and presentation packages are examples of generic applications software.

Word-processing package

Word-processing packages, such as Microsoft Word, are used to produce documents such as letters and reports.

Word-processing packages usually have the following features:

- Importing files from other software
- Typing text
- Formatting text, e.g., by using different fonts, sizes, colour, bold or italics
- Using bullet points to highlight important information
- Copying and pasting text
- Finding and replacing text
- Spelling and grammar checkers
- Inserting images
- Wrapping text around imported images
- Mail merge
- Displaying text in tables and columns

This book was written by two authors using a word-processing package. When one completed a chapter, he sent it to the other, as an email attachment, for checking and commenting. The word-processing package has a facility called "Track Changes" to show all the changes made by the reviewer. Deleted text is highlighted and added text is displayed in another colour. Comments can be added as well. When the reviewed chapter is returned to the author he can accept or reject each change.

Database management package

Database management packages, such as Microsoft Access, are used for information storage and retrieval.

A college could use a database management package to set up and access details of students and courses at the college. The package could be used to link students to courses. The data could be sorted so that class lists could be produced in alphabetical order. The information relating to an individual student could be retrieved through the use of a query.

Database management packages usually have the following features:

- Storing data
- Sorting data
- Searching a set of records
- Presenting information in reports
- Wizards to help the user set up the database
- Importing data from other packages

Spreadsheet package

Spreadsheet packages, such as Microsoft Excel, are ideal for storing, calculating and displaying financial information such as cash-flow forecasts, balance sheets and accounts.

Once the user has defined the cells and entered the formulas, if the contents of one cell are changed the contents of related cells and any graphs based on this data are changed automatically.

Spreadsheet packages usually have the following features:

- Storing data in a grid of rows and columns
- Calculations
- Mathematical functions
- Macro facilities to perform common tasks
- Copying and pasting data
- Finding and replacing data

Presentation software

Presentation software, such as Microsoft PowerPoint, might be used by the head of a sales team who wishes to present information during talks that he has arranged to give to groups of salesmen in different parts of the country before the launch of a new range of products.

Presentation software is increasingly used in the classroom. It usually has the following features:

- Creating slides
- Including text and graphics
- Formatting text, e.g. by using different fonts, sizes, colour, bold or italics
- Using bullet points to highlight important information
- Animated displays to attract the audience's attention and show one part of a page at a time
- Sound files for extra effect
- Video and animation
- Importing from other packages, for example, a sales presentation could import graphs of projected sales taken from a spreadsheet

Presentation software is normally used with a data projector for presenting to an audience. In the past this would have been done with an overhead projector (OHP). Presentation software is easier to use:

- Easier to edit
- Can't get slide order muddled
- Can include animation or video
- Can be automated and require no physical intervention.

▼

Web browser

Web browsers, such as Microsoft Internet Explorer or Netscape Navigator, are used to access websites. A web browser usually has the following features:

- Accessing the Internet
- Displaying a web page when a user types in the page's Uniform Resource Locator (URL) – the unique address
- Storing the addresses of "favourite" pages so that they can be accessed when required without the user having to remember the URL
- Storing the details of pages visited in a "history" folder
- Saving pages for viewing offline
- Storing previously loaded pages on the hard drive to reduce loading time if the user decides to go back to a recently viewed page
- Providing a link to a search engine

Web development software

Web development software, such as Adobe DreamWeaver, is used to develop new web pages. Web development software usually has the following features:

- Setting up and testing web pages
- Uploading pages to the Web
- Entering and formatting text
- Inserting tables
- Inserting hyperlinks
- Inserting and resizing images on a web page
- Displaying the HTML code for a page

Email software

Email software, such as Microsoft Outlook Express, is used to send emails and attachments. Email software usually has the following features:

- Sending emails to one or more addresses
- Setting up groups of users to send an email to
- A reply button for easy response
- Sending files as attachments
- Forwarding an email to another recipient
- Setting the priority of an email, e.g. Urgent
- Blocking emails from a particular sender

Desktop-publishing (DTP) software

Desktop-publishing software, such as Microsoft Publisher, is used to design and create printed documents, such as:

- Magazine pages
- Flyers
- Posters
- Invitations
- Leaflets
- Advertisements

A DTP package usually has the following features:

- Setting the page size
- Loading sample templates
- Formatting text, e.g. by using different fonts, sizes, colour, bold or italics
- Using bullet points to highlight important information
- Copying and pasting
- Finding and replacing
- Spelling and grammar checkers
- Inserting images

Image-manipulation software

Image-manipulation software, such as Adobe Photoshop, is used to prepare images, e.g. for publishing in magazines, newspapers or on the Internet. An image-manipulation package usually has the following features:

- Resizing images
- Cropping images
- Increasing or decreasing brightness
- Increasing or decreasing contrast
- Blurring or sharpening images
- Recolouring images
- Cloning (copying) part of an image
- Storing different parts of an image in layers

Activity 1

Listed below are a number of advanced features of a word-processing package. Check that you can use each of these features in your word processor, using the online help if you need to.

Feature	Example of use
Headers and footers	Customised text can appear at the top and bottom of every page in a document. Can include page number, date and time. The word processor automatically keeps track of page numbers so that the correct number appears on each page. (In a large document, such as a book, a different header can be used for each chapter.)
Creating a table of contents	Page numbers for chapter and sections can be inserted automatically into a table of contents.
Creating an index	An alphabetic list of terms linked to all references in a text can be created automatically.
Importing a graphic and wrapping text around it	A graphical image can be embedded in a document and the text set to flow around it.
Using templates	A template is a standard document that can be adapted to produce similar documents. It may include font styles, headers and footers, standard text and language used.
Mail merge	Mail merge is a process that merges text from one file with data from another. This is particularly useful for generating many files that have the same format but different data. For example, a standard letter can be created with a number of merge fields, such as the recipient's name, which are copied from the other file. Each output letter has a different name added to it.
Footnotes and endnotes	The numbering and placement of footnotes can be automated to enable easy cross-referencing.
Customised dictionary for use in spell checking	A spell checker is a utility within the word processor that checks the spelling of words. It highlights any words that it does not recognise. It is possible to create an additional dictionary with extra words that are commonly used by the user.
Search and replace	The word processor can search for a particular word or phrase. The user can direct the word processor to replace the word with another, everywhere that the first word appears.
Thesaurus	A built-in thesaurus allows the user to search for synonyms.

▶ Worked exam question

Word-processing packages can be used for many different tasks. Describe a task that you have completed using a word-processing package and explain how the functionality of the package helped you.　(4)

▶ EXAMINER'S GUIDANCE

The question carries four possible marks, two for describing the task and two for explaining how it helped. As with many ICT exam questions, it is quite easy to get one or even two of the marks but is harder to write an answer that will be awarded full marks.

Under the pressure of an examination, your mind might go blank and it is not easy to come up with a task. Think what you have used a word-processing package to do. For example, this might be:

- *to write a letter to apply for a part-time job*
- *to write a CV*
- *to produce a piece of homework*
- *to type out a recipe for your gran.*

A possible answer to the first part is:

▶ **SAMPLE ANSWER**   I used a word-processing package to write my CV.

▶ **EXAMINER'S GUIDANCE**   *This would only get one mark. It says what you used the word-processing package for without describing the task. A better answer is below:*

▶ **SAMPLE ANSWER**   I used a word-processing package to write my CV, typing in my contact details, my exam results, my employment history and my personal interests and hobbies.

▶ **EXAMINER'S GUIDANCE**   *In the second part, you need to explain how one feature of the software helped you. An answer such as the one below will get you no marks:*

▶ **SAMPLE ANSWER**   It was quicker.

▶ **EXAMINER'S GUIDANCE**   *What was quicker? Quicker than what? You need to state the feature and explain how it helped you. A better answer would be:*

▶ **SAMPLE ANSWER**   I need to update my CV regularly. It is easy to edit text using a word-processing package so I can make changes to my CV without having to type it all out again.

Activity 2

Now try to explain why some other features of a word-processing package might be useful in creating a CV such as:

- ■ spell checker
- ■ bold type
- ■ print preview.

Activity 3

Describe three advanced features of a word processor that would be useful to a solicitor's secretary, explaining in your answer how each feature would be used.

Activity 4

Find out about each of the following features commonly found in a spreadsheet package and give an example of the use of each one:

Feature	Example of use
Charts	Numeric data from the spreadsheet cells can be represented in many different graph or chart formats so that the data is easier to interpret.
Specialist functions such as IF, VLOOKUP	
Pivot table	
Linked multiple sheets	
What-If modelling	

▶ Specific applications software

Specific applications software is designed to carry out a particular end-user task, usually for a particular industry. It is of little use in other situations. The user will normally require knowledge of the subject area involved. A payroll application program would be an example of specific applications software; it is designed to be used exclusively for payroll activities and could not be used for any other tasks.

Specific applications software is available for wide-ranging areas such as engineering and scientific work; there is a range of specialist design tools including computer-aided design (CAD) packages and sophisticated mathematical software.

Simple graphics software, such as Paint, which comes free with Windows, stores graphics files in bitmap form. This stores the colour of each pixel (dot) in the picture. As a result, bitmaps are large files and, when they are resized, the image tends to be distorted.

Sophisticated graphics packages, such as CorelDraw, store graphics files in vector graphic form. This means it stores pictures as a series of objects such as lines, arcs, text, etc. Facilities in the software allow individual objects to be resized, rotated and have fill colour and line characteristics changed. When storing a line, it stores the co-ordinates of the start and finish of the line, its width and colour. Not only does this save space, it means that the line can be moved, deleted or changed in colour or width without affecting the rest of the picture.

CAD programs are used for designing, for example, in engineering or architecture. Users can draw accurate straight lines and arcs of different types and thickness. By zooming in, designs can be produced more accurately. Designs can be produced in layers to show different information. For example, one layer might show electrical wiring and another gas pipes.

Special-purpose CAD packages are available for many areas of design such as kitchen or garden design.

Software designed for use in doctors' or dentists' surgeries and booking systems for leisure centres or cinemas are all available.

Which software should I use?

Suppose you are purchasing a new word-processing package. There are many programs available. How do you decide which one to buy? You are likely to use a variety of criteria in your decision:

- Performance – How quickly does it operate?
- Functionality – Does it have all the features I want?
- Usability – Is it easy to use?
- The interface – Is the screen layout clear and not too cluttered?
- Compatibility – Can the software run on my existing computers?
- Transferability of data – Can it load files created in my existing word-processing software and files that other people are likely to send to me?
- Robustness – Is it reliable or does it crash?
- User support – Is telephone support available? Is it there 24/7? Does it cost extra?
- Hardware requirements – What processor speed, RAM and hard drive space are required?
- Software requirements – Will it work with my version of Microsoft Windows?
- Financial issues – What will it cost?
- Manufacturer's credibility – Does the manufacturer have a good reputation?

You are likely to give each of the criteria a maximum score depending on how important you think each one is. Then you give each package a score for each criterion. The total score for each package will help you decide.

You might also look at software reviews.

case study
▶ Microsoft Word or OpenOffice Writer?

Microsoft Word is the biggest selling word-processing package in the world. It is produced by Microsoft, the world's biggest software company. Usually purchased as part of the Microsoft Office package, it can also be bought on its own for around £80–£100.

OpenOffice Writer is a word-processing package that is part of a rival office suite called OpenOffice.org. The whole package can be downloaded for free from http://www.openoffice.org. It is completely free and it does work.

Here are one user's comments on OpenOffice Writer:

- **Performance** – I've heard some reports suggest that some features are not as fast as with Microsoft Office but I have had no problems.

- **Functionality** – "Neck and neck" with Microsoft Word on features.

- **Usability** – Easy to use; it is very similar to Microsoft Word.

- **Interface** – As you might expect, the screen layout is very similar to Microsoft Word.

- **Compatibility** – The software can run on existing computers.

- **Transferability of data** – Writer can read all your old Microsoft Word documents and save your work in Microsoft Word format.

- **User support** – There is free community support, discussion groups, help, FAQs, etc.

- **Software requirements** – OpenOffice runs on Windows or Linux.

- **Financial issues** – It's completely free. Could you have a better price?

- **Manufacturer's credibility** – OpenOffice's reputation is improving all the time.

Activity 5

1. If you are creating web pages, three possible software packages to use are Microsoft FrontPage, Expression Web and Adobe DreamWeaver. Use the Internet to investigate the advantages and disadvantages of each package. Hint: A good starting point is to go to a search engine and search for "DreamWeaver FrontPage".

2. If you are creating a spreadsheet, you could use Microsoft Excel or OpenOffice Calc. Use the Internet to investigate advantages and disadvantages of each package.

Software is the name given to computer programs, without which no computer hardware can function. There are two main types of software:

▶ Systems software is programs which help the computer run more smoothly. Utility programs and operating systems are types of systems software. Examples of utilities include programs to delete, rename or copy files, format disks and compress files.

▶ Applications software is programs that carry out an end-user task, such as calculating staff wages. Applications software can be:

    ▶ generic programs that can be used by many users in a variety of ways.

    ▶ special-purpose software written to meet a particular need, such as a booking system or a design package.

Common examples of generic applications software are word-processing software, spreadsheet software, database software and browsers (software that allows you to access the information stored on the Internet).

You are likely to use some of the following criteria in your decision of which software to buy:

▶ Performance

▶ Functionality

▶ Usability

▶ The interface

▶ Compatibility

▶ Transferability of data

▶ Robustness

▶ User support

▶ Hardware requirements

▶ Software requirements

▶ Financial issues

▶ Manufacturer's credibility

Questions

◀

1. Spreadsheet packages can be used to solve many different problems. Describe a problem that you have solved using a spreadsheet and explain how the functionality of the spreadsheet helped you to solve that problem. (6)

   AQA Specimen Paper 1

2. State three formatting facilities that are offered by word-processing software. (One word answers are acceptable for this question.) (3)

   June 2005 ICT2

3. Word-processing software offers editing and formatting facilities.
   a) Explain the differences between formatting and editing. (2)
   b) Describe three editing facilities that are offered by word-processing software. (6)
   c) Describe three formatting facilities that are offered by word-processing software. (6)

4. A map has been scanned and its image saved in a file. State three ways in which the image of the map could be manipulated before it is printed. (3)
   (The use of brand names will not gain credit.)

   January 2003 ICT2

5. Personal computer systems are usually supplied with some system software already installed. Explain, using examples, the purpose of system software. (4)

   June 2003 ICT2

6. Software packages contain many advanced features that are often used by certain people. Explain, with the use of examples:
   a) two advanced features of a word-processing package that would be useful to an author writing an ICT textbook (6)
   b) two advanced features of a spreadsheet package that would be useful to an accountant preparing a financial report. (6)

   June 2003 ICT2

7. A company has 12 hotels in locations in the north of England. It has decided to upgrade its ICT systems. At present, it only uses its computers for word processing. Name and describe four new items of software that the hotel could buy. Explain why each item would be required. (12)

8. Every month, the organiser of a dog training group, Mrs Jones, sends an email to all the members in her area. The email has been sent to Mrs Jones from the National Dog Training Centre. She likes to add a scanned picture of "dog of the month" to the email before she sends it on. State three functions provided by email software and explain how each one would help her to carry out the above task efficiently. (6)

   AQA Specimen Paper 1

8

Implementation of ICT-related solutions

AQA Unit 1 Section 8

In this section, you have to set up a working solution which meets the stated client requirements.

You need to provide evidence that your solution meets all the requirements. The evidence of your solution is likely to include:

- Annotated printouts from the solution
- Annotated screenshots from the solution
- Witness statements from a teacher that they have seen your solution working.

In Chapter 9, you learn how to test that your system works.

Website solutions

Go back to the original input, processing and output requirements (see Chapter 2). As you create your web pages, check that the requirements have been met.

example
▶ **output requirements**

There were 12 output requirements for the web pages, as follows:

1. A home page linked to pages for each country

2. List and flags of the four countries

3. Thumbnail images of the cottages

4. Maps

5. Links to a page for each cottage

6. Images of the cottages

7. Information about the cottages – the cottage code, the cottage name, the village, the region, the price for one or two weeks, high or low season and other important details

8. Details of forthcoming shows including dates and the venue

9. Contact details

10. Details of the company

11. The company name and logo

12. Links to other pages

example (contd)

How I have met these output requirements

1. A home page linked to pages for each country: the home page is shown in Figure 8.1. The flags are hyperlinks to the pages for each country: France, Spain, Italy and if you scroll down, Germany.

Figure 8.1   Home page

2. List of the four countries and their flags: the flags of the four countries and their names are on the home page.

3. Thumbnail images of the cottages: there are thumbnail images of each cottage on the country page as shown in Figure 8.2. The thumbnails are hyperlinks to details of the cottages.

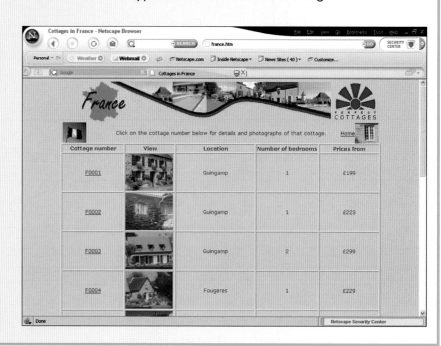

Figure 8.2   Country page

example (contd)

4. Maps: there is a map of how to find each cottage on the cottage page as shown in Figure 8.3.

5. Links to a page for each cottage: as you can see in Figure 8.3, there is a page for each cottage. The link is from the thumbnail.

6. Images of the cottages: there is one large photo of each cottage, as in Figure 8.3.

7. Information about the cottages: the cottage code, the cottage name, the village, the region, the price for one or two weeks, high or low season and other important details are all included in the page for each cottage (see Figure 8.3).

Figure 8.3   Cottage page

8. Details of forthcoming shows including dates and the venue: there is a link to this information at the bottom of the home page (see Figure 8.4).

9. Contact details: a hyperlink to full details is at the bottom of the home page (see Figure 8.4).

10. Details of the company are on the home page.

11. The company name and logo are on every page.

example (contd)

12. Links to other pages: each cottage page has a link to the home page and the page for the country it is in.

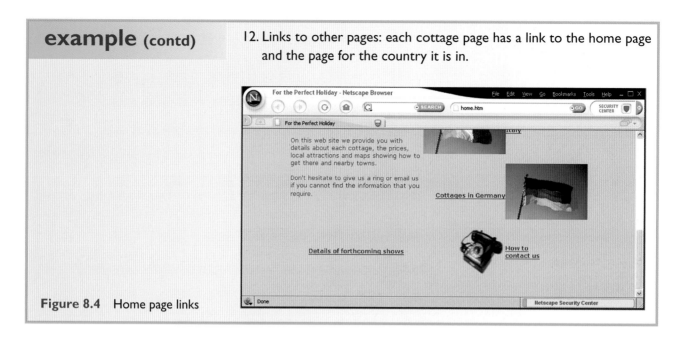

Figure 8.4　Home page links

You need to go back to the original input and processing requirements and check that these too have been met, providing evidence.

<hints>

■ You should be able to explain what any HTML script you have written does.

■ If you have set up a style sheet, explain how it meets the client requirements and include a screenshot (see Figure 8.5).

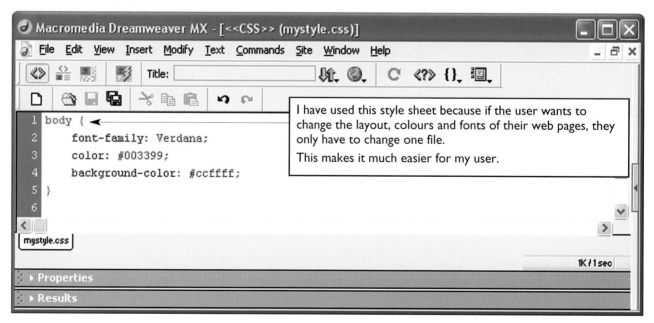

Figure 8.5　Style sheet

Spreadsheet solutions

If you are creating a spreadsheet solution, you need to build the solution and show that you understand what it does and how it meets the client's requirements.

In the same way as for a website solution, you need to refer to the original requirements and provide evidence through screenshots and printouts.

An example of meeting the input requirements from the spreadsheet solution in Chapter 2 is given below.

example

▶ **input requirements**

There were four input requirements for the spreadsheet as follows:

- the type of paper or card

- the quality of the paper or card

- the quantity required

- whether it is folded or not.

How I have met these input requirements (Figure 8.6)

- The type of paper or card: I have set up a combo box on the quotation sheet so that the thickness of the paper or card can be chosen easily.

- The quality of the paper or card: I have set up option buttons on the quotation sheet so that the quality of the paper or card can be chosen easily.

- The quantity required: This is entered in cell D9 and can be adjusted easily using the spinner which goes up and down in hundreds.

example (contd)

■ Whether it is folded or not: A check box is used to indicate whether the printed material should be folded.

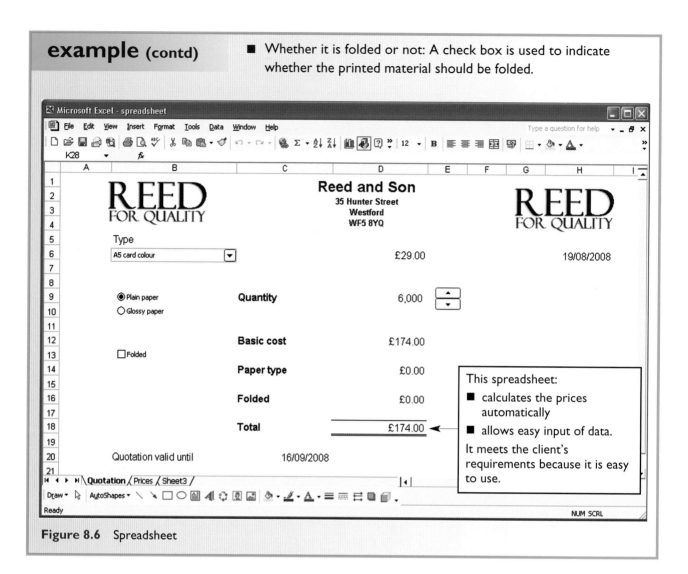

Figure 8.6 Spreadsheet

Make sure that you annotate your screenshots and printouts to point out how your spreadsheet meets the client's requirements.

You may need to provide annotated screenshots of formulas and the coding of macros as evidence of meeting requirements. You should be able to explain the formulas that you have set up and how they meet the client's requirements

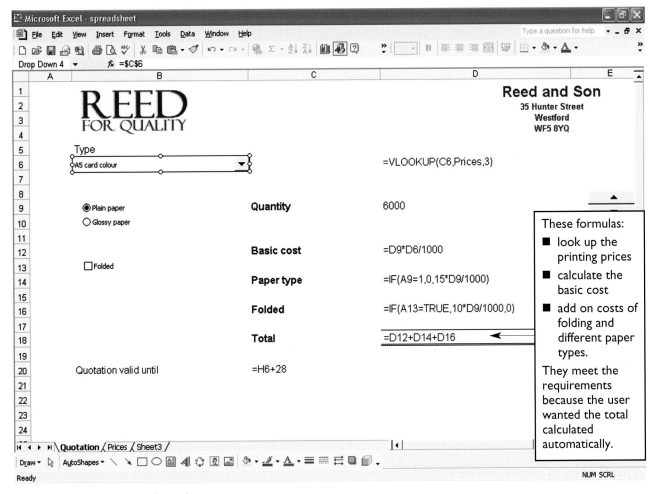

Figure 8.7 Spreadsheet formulas

<hints>

■ Explain how you have met each requirement.

■ Provide annotated screen shots where required.

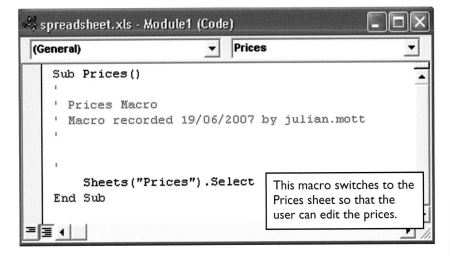

Figure 8.8 Spreadsheet macro

In the examination you may need to explain what a formula or a macro does, for example:

- This formula adds the basic cost of the printing to the additional cost of glossy paper and folded paper.
- This macro switches to the Prices sheet so that the prices can be edited.

Remember you are trying to show that you have met your client's requirements. You don't need to include screenshots of the in-between stages in setting up the solution. Only include evidence of the finished solution.

9

Testing of ICT-related solutions

AQA Unit 1 Section 9

Test plan

Having created your solution, you need to test it fully before you can give it to the client. Firstly you need to draw up a test plan that will test all the aspects of the solution in a logical order.

The test plan consists of a series of tests which, if successful, will ensure that your solution is working properly.

Test plans should include a range of suitable test data together with expected outcomes. Your test plan should include tests that ensure:

- the output is 100 per cent accurate
- the output is clear
- data input is validated
- printed output is as expected, for example, it all fits on one side of A4
- that the solution meets the requirements of the client
- that the solution is useable by the end user or intended audience.

One of the objectives of testing is to provoke failure – to try to make something go wrong. If you have tested all aspects of the solution and have been unable to provoke failure, you can be confident that the system is fully working.

▶ Testing against the plan

Once the test plan has been created, you need to follow your test plan in a systematic way to test the solution. You must provide evidence of each test in the form of screenshots and printouts, or witness statements if hard copy evidence cannot be shown.

Where the results of a test don't match the expected outcome, corrective action needs to be taken, the test repeated and evidence provided of it working.

The results of testing are used in the evaluation of the solution (see Chapter 10).

▶ Website solution test plan

Your test plan will need to include:

1. testing the web pages offline
2. uploading pages and images to the Internet
3. testing the web pages online
4. user testing to make sure that your user can use the solution.

You are not just testing that the hyperlinks work. You also need to test that:

- pages are readable – that text is clear against the background
- all images are present
- alternative text for images is working (if used)
- images are not pixelated
- load times are acceptable
- if a style sheet exists, the pages are in that style
- any rollovers work
- any hotspots work
- any counters increase by one
- any animation works
- any dynamic content works
- that the user can edit cottage details
- that a member of the potential audience of the website finds it easy to use
- that the client's requirements have been met
- that your client is happy with the web pages.

- Do not have dozens of hyperlink tests. Include just one test that tests all hyperlinks.

- When you come to test the online version use a different computer – one where the files are not stored on the hard drive. You may be loading images and pages from your computer's hard drive and not from the Internet.

- When considering load times, do not say it must load quickly. You must be able to measure the success of the test. Say it will load in three seconds or less.

One way of producing the test plan is to produce a table. Make sure that you number the tests. The actual outcome column is blank at this point. The evidence column will allow you to cross-reference to the evidence, i.e. say on which page the evidence can be found.

example

▶ part of a test plan

Offline tests

	Expected outcome	Actual outcome	Evidence
1. Test that all the hyperlinks work	All hyperlinks work satisfactorily		
2. Test that all images load as expected	All images load satisfactorily		
3. Test that style sheets load correctly (styles to fonts, sizes, colours)	Text is navy blue, Verdana; background is light blue		
4. Test that all text is clear and readable	Text is big enough to read and provides a good contrast		
5. Test that all images are clear	Images should not be pixelated		
6. Check the spelling on every page	There should be no spelling mistakes		
7. Test that all mouseovers work	Every image goes darker when the cursor moves over it		
8. Test the web pages in a number of different browsers and screen resolutions	Pages load and work with any browser software.		

Online tests

	Expected outcome	Actual outcome	Evidence
1. Test that the URL loads as expected	The website loads when the URL is typed in		
2. Test that all the hyperlinks work	All hyperlinks work satisfactorily		
3. Test that all images load as expected	All images load satisfactorily		
4. Test that load times are acceptable	Every page loads in less than 5 seconds		
5. User testing with Jodie and Monica	Users able to edit pages and upload pages and images		
6. User testing with Gerry Collier, potential member of audience	Use of website is intuitive		

▶ Spreadsheet solution test plan

Your test plan will need to include:

- testing the validation of data input
- testing individual outputs
- testing the total output
- testing that the client's requirements have been met
- sets of test data.

You are not just testing the total output. You also need to test that:

- the worksheets are readable – that text is clear against the background
- the worksheets are not too cluttered
- any macro buttons and hyperlinks work
- your client is happy with the spreadsheet.

\<hints\>

- Do not have dozens of button tests. Include just one test that tests all buttons.

- Test the total output several times with different sets of test data.

- The expected outcome should specify exactly what the total cost will be for each data set.

Again one way of producing the test plan is to produce a table. Make sure that you number the tests. The actual outcome column is blank at this point. The evidence column allows you to cross-reference to the evidence, i.e. to say on which page the evidence can be found.

example
▶ test data

Data set 1: A4 black and white – two sides. 6000 copies. Glossy paper. Folded.
Data set 2: A3 colour – one side. 4000 copies. Plain paper. Not folded.
Data set 3: A5 card colour. 10,000 copies. Plain paper. Not folded.

example
▶ test plan

Test	Expected outcome	Actual outcome	Evidence
1. Test that all the macro buttons work	All macro buttons work satisfactorily		
2. Test that all text is clear and readable and the page is not too cluttered	All text is big enough to read and provides a good contrast		
3. Test that all images are clear	The logo and the company name are clear		
4. Check the spelling of every worksheet	There should be no spelling mistakes		
5. Test VLOOKUP. Choose A4 colour – two sides	Price per 1000 should be £32.00		
6. Test spinner	Quantity should go up and down in 100s; lowest number is 1000		
7. Test Basic Cost: Data set 1	£168.00		
8. Test Paper Cost: Data set 1	£90.00		
9. Test Folded: Data set 1	£60.00		
10. Test Total cost. Data set 1	£318.00		
11. Test Total cost. Data set 2	£220.00		
12. Test Total cost. Data set 3	£174.00		
13. Test protection of formulas	You cannot edit the formulas		
14. Test validation of the quantity	1000 (extreme value) – OK 990 (erroneous value) – rejected		
15. User testing with data sets 1, 2 and 3	As for tests 10 to 12		

Once you have drawn up your test plan, follow it in order and correct any mistakes that you find in the solution.

Test reports

◀

Document the testing by copying the test plan and filling in the actual outcomes.

Include screenshot evidence immediately underneath the table or include a reference to where to find the evidence. Here are part of the test reports for a website and a spreadsheet solution.

example
▶ **website solution test report**

Offline test

		Expected outcome	Actual outcome	Evidence
8.	Test the web pages in a number of different browsers and screen resolutions	Pages load and work with any browser software	Pages worked in the following browsers: ■ Netscape ■ Internet Explorer ■ Firefox Pages were clear in 800 by 600 and 1024 by 768 resolutions but it was necessary to scroll across and down to see the image of the cottage	Testing Screenshots 7 to 11

Online test

5.	User testing with Jodie and Monica	Users able to edit pages and upload pages and images	Both Jodie and Monica were able to edit pages and upload edited pages and new images	Witness statement 1

example (contd)

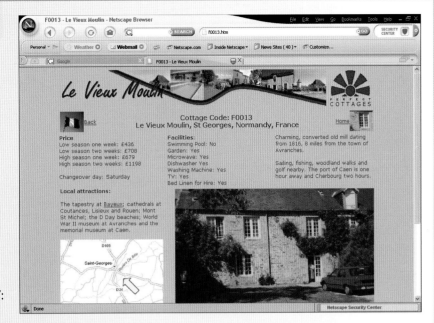

Figure 9.1 Testing Screenshot 7: Evidence for Test 8 (Netscape)

Figure 9.2 Testing Screenshot 8: Evidence for Test 8 (Internet Explorer)

example (contd)

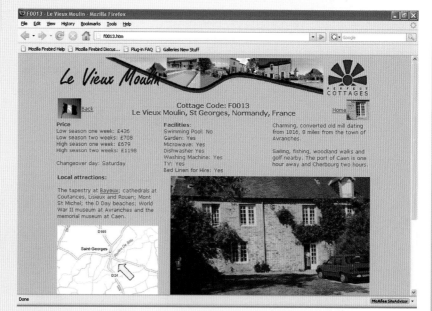

Figure 9.3 Testing Screenshot 9: Evidence for Test 8 (Mozilla Firefox)

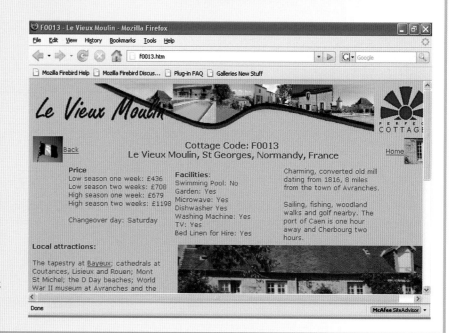

Figure 9.4 Testing Screenshot 10: Evidence for Test 8 (800 by 600 resolution)

example (contd)

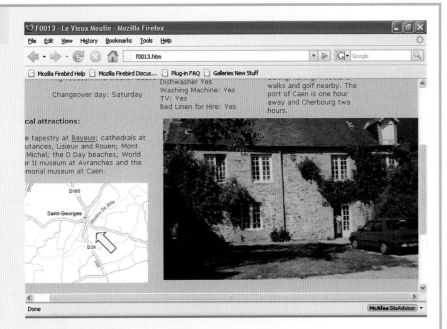

Figure 9.5 Testing Screenshot 11: Evidence for Test 8 (800 by 600 resolution scrolled to image)

example
► **witness statement 1**

I was able to do the following:

- Edit the price of a cottage on a page

- Replace a photo with another image

- Upload the replacement image

- Upload the replacement page

- Test that the edited version was working.

Monica Holden 19/11/07

example
▶ **spreadsheet solution**
test report

		Expected outcome	Actual outcome	Evidence
7.	Test Basic Cost: Data set 1	£168.00	£168.00	Testing Screenshot 5
8.	Test Paper Cost: Data set 1	£90.00	£90.00	Testing Screenshot 5
9.	Test Folded: Data set 1	£60.00	£60.00	Testing Screenshot 5
10.	Test Total cost: Data set 1	£318.00	£318.00	Testing Screenshot 5
13.	Test protection of formulas	You cannot edit the formula in cell D12	Error message appears; you cannot edit the formulas	Testing Screenshot 12
14.	Test validation of the quantity	1000 (extreme value) – OK 990 (erroneous value) – rejected	1000 – OK 990 – error message appears	Testing Screenshot 13

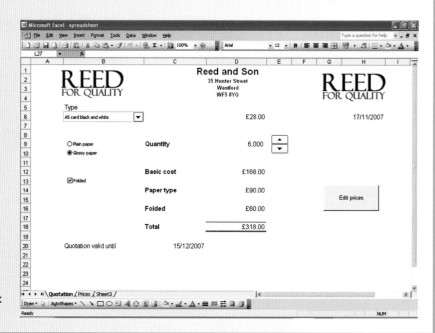

Figure 9.6 Testing Screenshot 5: Evidence for Tests 7 to 10

example (contd)

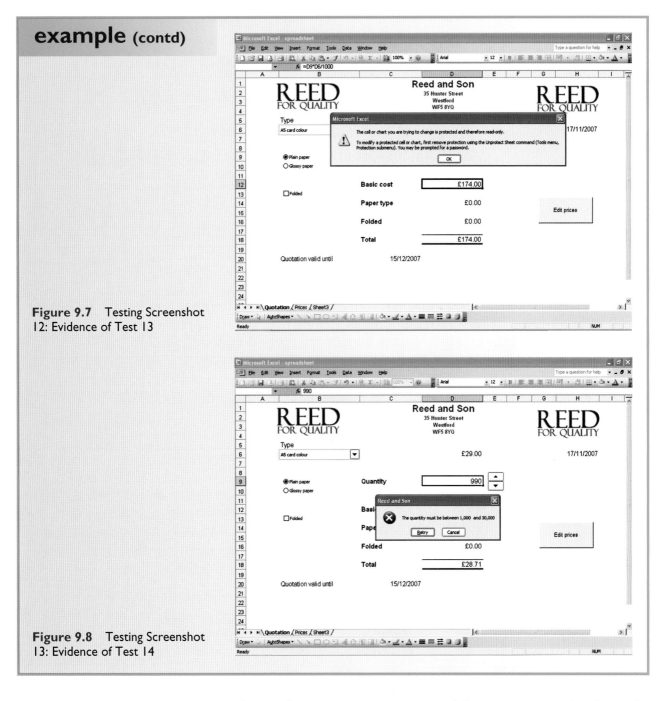

Figure 9.7 Testing Screenshot 12: Evidence of Test 13

Figure 9.8 Testing Screenshot 13: Evidence of Test 14

If any of your tests are unsuccessful, correct your mistake and perform the test again.

<hints>

■ Do make sure that screenshots are not so small that it is impossible to read them.

■ If you print out in black and white, all details must be clear.

■ If you crop any screenshots, do not cut off any part of the image that may be useful evidence. For example if there is an important formula in a spreadsheet, do not cut off the formula bar.

It is important to evaluate your solution. Does it do what you were originally asked to do? Does it do it how it was supposed to do it? How well does it do it? Are there any limitations? Could it be (even) better?

Does it do what you were originally asked to do?

The first thing is to check that the solution meets the client's requirements

Go back to Chapter 2 and find your original client requirements. For each requirement say whether you have done it or not. Refer back to the results of your testing. You may wish to provide evidence in the form of screenshots or the results of tests.

---

example
▶ does the solution do what it was supposed to do?

My client requirements were as follows:

1. An easy to update website – it is easy to update the web site by editing the web pages offline and then uploading the new pages following the instructions in the manual. Both Jodie and her assistant Monica have successfully edited pages and uploaded pages and images.

2. Intuitive for the audience – The website uses flags and images of cottages. These images are easy to recognise and so are intuitive. The cottages are listed in cottage number order which makes them easy to find.

3. Quick to load – I have loaded the home page on my home computer using ADSL broadband and at school. Because there are only small images, on both occasions it took less than 2 seconds to load which is very fast.

4. Modern image – I think that my designs are modern. They are very clean and easy to read.

example (contd)

5. Company logo on each page – the company logo is on every page in the top right-hand corner see Figures 10.1 and 10.3.

6. Verdana font – I have used it throughout, see Figures 10.1 and 10.4.

7. Pale blue background but with navy and red flashes – I have implemented a pale background with navy and red flashes, see Figures 10.1 and 10.3. The text is navy blue. A cascading style sheet has been used to set the colours. By changing one file, I can change all the web pages.

8. Links to other pages and photos – There are links to all other pages, see Figures 10.1 and 10.3.

9. Thumbnail images as links – The website uses flags and thumbnail images of cottages throughout as hyperlinks (see Figure 10.1).

10. Mouseovers for hyperlinks – Every hyperlink images uses mouseovers. As the cursor moves over the picture, the picture goes darker. The mouseovers have been tested and are fully working. Figure 10.2(a) shows the France hyperlink on the home page. Figure 10.2(b) shows that the image goes darker when the mouse moves over it.

11. Thumbnails and photos of all cottages (141 in total) – Photos of many of the cottages are on the website, see Figure 10.1.

12. Details of all the cottages – These are included, see Figure 10.3.

13. A page for each cottage with photos taking up about half a page and a map – These are completed for several cottages, see Figure 10.3.

14. Contact details – These details have been included.

15. Details of the company – These details have been included.

16. Full updating instructions – Full instructions have been produced and supplied to the client. Both Jodie and Monica have used these instructions.

17. Ability to update prices, delete cottages, add new cottages – This can be done easily. Both Jodie and her assistant Monica have successfully edited pages and uploaded pages and images (see online test 5 and witness statement 1).

18. 100% fully working in any browser – The website has been tested in Netscape (see Figures 10.1 and 10.3) and also in Internet Explorer (Figure 10.4) and Firefox (Figure 10.5). All tests have been successful and I believe the site to be fully working.

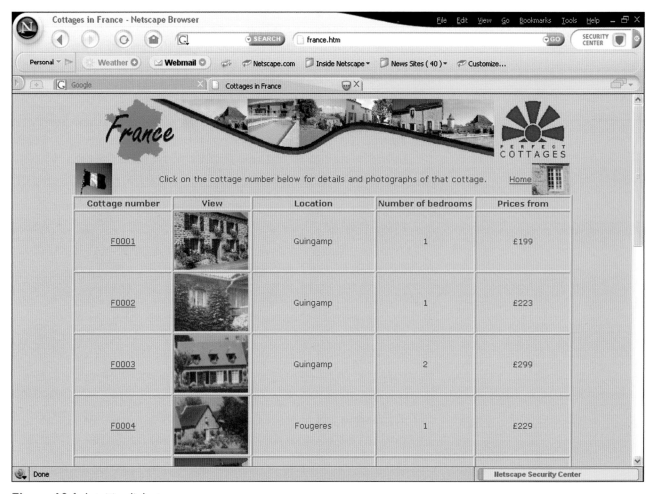

Figure 10.1 Intuitive links to pages

Figure 10.2 Image hyperlink a) when the mouse is not over it b) when the mouse is over it

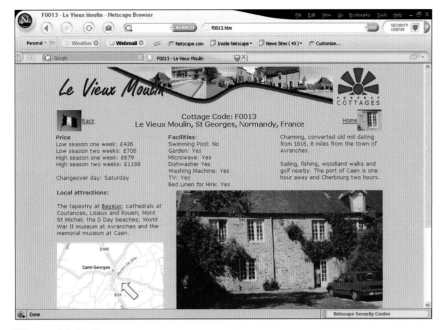

Figure 10.3 Cottage page

Figure 10.4 Home page in Internet Explorer

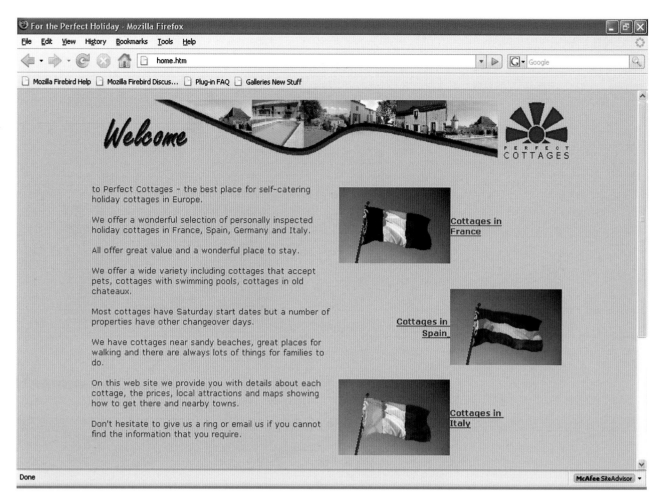

Figure 10.5 Home page in Mozilla
Firefox

Is the solution an effective one? ◄

To assess whether your solution is effective, you need to
consider evaluation criteria. These criteria look at whether the
system does everything it should, whether it is fully working
and how well it performs.

Evaluation criteria are likely to include maximum load
times and file sizes – particularly if you have set up web pages
where the load times are important.

Possible evaluation criteria include:

- Does it do everything the client wanted?
- Is it fully working?
- Do all pages load in less than 5 seconds over a dial-up
  connection?
- Do all pages load in less than 2 seconds over a broadband
  connection?
- Are all images no greater than 20 kB, unless they are
  loaded from a thumbnail of the image?

example
▶ **evaluation criteria**

1. As you can see, all the client requirements have been met.

2. All parts of the solution have been successfully tested. All pages work now that they have been uploaded to the Internet.

3. I have tested all the pages using a dial-up connection. They took between 1 and 4 seconds to load so they all meet the criterion that they load in less than 5 seconds.

4. I have tested all the pages using a broadband connection. They all took less than 1 second to load so they all meet the criterion that they load in less than 2 seconds.

5. All thumbnail images are less than 20 kB in size (see Figure 10.6) and so load quickly.

6. The Perfect banner is 14 kB (see Figure 10.7) and so is less than 20 kB in size and can load quickly.

7. The original images can be as big as hundreds of kilobytes (see Figure 10.8) so thumbnails are used.

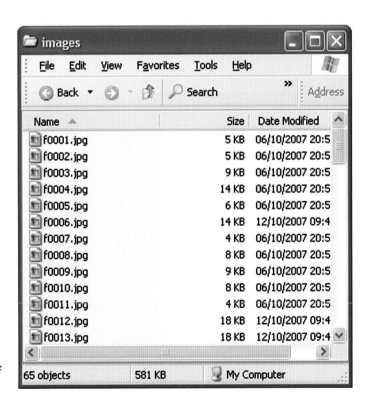

Figure 10.6 Detailed folder listing of thumbnail images

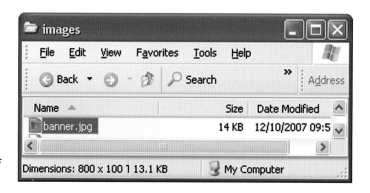

Figure 10.7 Detailed folder listing of banner image

Figure 10.8 Detailed folder listing of original image

Improvements

Write down some ways in which you could improve your solution.

example

▶ **potential improvement to my solution**

On reflection, I feel that I could make my website better. The banner is a good idea but it could be clearer. There is also a lot of unused space on the home page. There is also a lot of text which I don't think people would read.

If I started all over again, I would:

- Redesign the banner to improve the images. At present too much space is devoted to swimming pools and the audience cannot see what the cottage might look like.

- Discuss with my client (Jodie) and members of the potential audience how well the cottage pages display the facilities available. At the moment they tend to be lost in the middle of other text. We could consider making this information stand out by using a bigger font or maybe a different colour.

- Consider having pictures of the inside of each cottage as well as the outside.

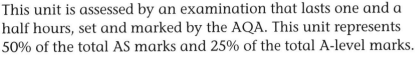

Preparing for Unit 1 exam

AQA Unit 1 Exam preparation

This unit is assessed by an examination that lasts one and a half hours, set and marked by the AQA. This unit represents 50% of the total AS marks and 25% of the total A-level marks.

The exam is in two sections:

- Section A has short answer questions.
- Section B has longer, more structured questions.

You must answer all the questions.

Sample work

You must take into the exam your sample work – that is, the documentation from the two problems for which you have created a solution.

In particular, the examiner will be looking at:

- How you identified the problem, your client's requirements and how you have broken these requirements down into input, processing and output.
- Your test plan, your test report and evidence (that is screenshots and printouts) of your testing.

Your sample work must be between 10 and 20 pages of A4. It must be succinct, clearly laid out and legible to the examiner. Pages should be numbered with your name, candidate number and centre number in the header or footer.

The sample work should be fastened with a treasury tag in the top left corner. At the front should be fastened the Candidate Record Form which you must complete and sign.

The exam is not based solely on your project work. You will also get questions on the material in Chapters 1 to 10 of this book.

Before the exam

1. Complete your sample work.
2. Put your name, candidate number and centre number in the header or footer.
3. Number the pages.
4. Produce a contents page.

5. Fill in and sign the candidate record form.
6. Fasten the work together with a treasury tag so that it can easily be attached to the back of the exam paper.
7. Hand the work in to your teacher to sign and give back to you before the exam.
8. Revise the material in Chapters 1 to 10.
9. Look at the worked exam questions in Chapters 1 to 10.
10. Use the questions to practise exam style answers using your sample work where asked to.
11. Prepare for questions on your sample work by answering the exam preparation questions on page 97.
12. Practise other questions.

Exam preparation questions ◀

1. Who is your client?

2. Who will be the users of your solution?

3. Copy one of your client's requirements. (1)

4. State one input involved with this client requirement. (1)

5. State one process involved with this client requirement. (1)

6. State one output involved with this client requirement. (1)

7. Explain how you have tested whether this requirement has been met or not. (2)

8. If you have set up a website solution, explain how you have tested the download times of your website. (4)

9. If you have set up a spreadsheet solution, explain how you have used boundary data to test your solution. (4)

10. What is the purpose of your test plan? (3)

An ICT system and its components

◀

What is information and communication technology?

◀

Information and communication technology (ICT) means using any form of **digital technology** for the **input**, **storage**, **processing** and **transferring** of data and the **output** of information.

The digital technology could be a computer or another device, such as an MP3 player, a mobile phone or a satellite television.

▶ What is a system?

Any activity or set of activities that involves input, processing and output is called a system. Examples in biology include the digestive and respiratory systems. The electoral system forms a major part of a democratic society.

▶ What is an ICT system?

An ICT system is a system that uses any form of digital technology; the output goes directly to a human being or into another ICT system. Examples of ICT systems include a text messaging system on a mobile phone, an invoicing (billing) system for a mail-order company and an online holiday booking system. An ICT system is the combination of **hardware** and **software**. A typical ICT system has programs (software) that convert the system's inputs into outputs.

Input, processing and output

◀

Input means entering data into the computer or other digital device. For example, the data input could be bank account numbers, the numbers identified by a bar code or the pressing of a key representing a selection of a menu choice on a mobile phone.

Processing means manipulating the input data into information in a form that is understandable and useful to the user. This might be by counting up the number of items purchased by a customer and adding up the amount of money owed.

Output means presenting this information to the user or the outside world. It must be in a form the user can understand and finds useful, so it must have a context. It could be printed, displayed on a screen or in another form. The output can be to a human being or to another ICT system.

Many ICT systems have a number of inputs and outputs. They may also include a number of processes.

example 1
▶ in a school or college

- Attendance might be collected and typed into an ICT system at each class (**input**).

- At the end of the week, calculations are made on the stored collection of attendance data. This might include summarising, adding up and sorting the data (**processing**).

- At the end of the week, tutor group (or form) lists are printed showing each student's overall attendance during the week. This is useful information for the tutor (**output**).

example 2
▶ photographs are taken using a digital camera

- The image is **input** as light waves.

- **Processing** converts the images into digital form.

- The digital image is **stored** on the camera's memory card.

- The image can be viewed on the camera's LCD screen (**useful information output to a human**) or can be transferred to a computer for editing (**output to another ICT system**).

Activity 1

Copy the grid below and fill in the missing cells.

Context	What data is input?	What is the processing? (there can be more than one form of processing)	What information is output? (there can be more than one output)	Where does the output go (to a human or another ICT system)?
Payroll	Employee name, hours worked	Calculate pay, work out tax, etc.	Payslip details	To the employee
Find phone number for mobile phone call				
Household electricity supply			Bill details	
Examination board	Marks for exam papers			
MP3 player			Musical sound	
Salesman's visit			List of appointments	
Going to a cash machine				
Booking a holiday online				

case study 1
▶ at the supermarket till

Figure 12.1 EPOS at a supermarket

The supermarket uses an ICT system for adding up the cost of a customer's purchases. Instead of personally adding up the amount, the checkout operator just has to make sure that data is entered into the ICT system using the EPOS (electronic point of sale) device. The software then carries out the necessary calculations. The system has a large number of inputs and outputs.

Products are passed over a flatbed, bar-code laser scanner which interprets the product identification code that is encoded in the bar codes. This code uniquely identifies the product. The price is found from a database stored as part of the ICT system on a hard disk that is probably located somewhere in the store. A specialised keypad allows numbers to be entered if several of the same items are purchased; this saves the operator from scanning every one.

Whenever an item is successfully scanned a beep sound is made (through a speaker). If, for some reason, the bar code cannot be read a different sound is made to alert the operator. The operator can then enter the product identity code, which is printed underneath the bar code, using the keypad. As products are scanned, a screen displays details of the latest purchase and the current total cost for the customer to view.

Loose products such as fruits and vegetables do not have a bar code. These items must be weighed on special scales which use sensors to produce an electronic value equivalent to the weight. The

case study (contd)

operator must then choose the product type from a series of images displayed on a touch screen.

A second, related, ICT system deals with customer payments made with a credit or debit card. "Chip and PIN" allows the customer's smart card to be read by a small device connected to the till. The customer enters a PIN using the device's keypad. The customer's bank account is accessed via a wide area network link to the bank and the transaction is authorised. At the till, a small, built-in printer is used to print out the bill which lists every item purchased together with its price. Details of the transaction are then sent via the wide area network to a computer at the customer's bank so that the account can be updated by the bank's ICT system.

1. Identify all the inputs to the EPOS system.

2. Identify all the outputs from the EPOS system and for each state whether it goes to a human (who?) or to another ICT system.

3. Identify all the processes in the EPOS system.

4. What data is stored in the EPOS system?

5. Identify all the inputs to the payment system.

6. Identify all the outputs from the payment system and for each state whether it goes to a human (who?) or to another ICT system.

7. Identify all the processes in the payment system.

8. What data is stored in the payment system?

Components of an ICT system ◀

An ICT system consists of the following components:

- Data
- Hardware
- Software
- Information
- Procedures
- People

▶ Data

Data are raw facts or figures, a set of values or measurements, or records of transactions. By itself, without a context, data has no useful meaning to a human user. For example, data can be the four digits representing the PIN associated with a customer's debit card or the characters representing the contents of a text message. See Chapter 13 for a detailed description of data.

► Hardware

Hardware is the name given to the physical components of the ICT system. A printer, a MP3 player and a hard disk drive are all examples of hardware.

ICT systems work on the basis of input, processing and output. In many systems, data is transmitted from one computer to another. The various parts of the computer (the hardware) can be defined as input devices, the processor, backing store, communication devices and output devices.

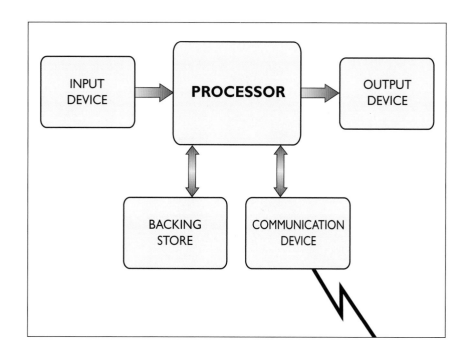

Figure 12.2 Parts of a computer system

Input devices are used to enter data. The processor is in the "box" part of the computer and is made up of electronic printed circuits and microchips. The processor includes the computer's main memory. Backing store is where data is stored when the computer is turned off. Output devices are used to present the information to the user.

The **central processing unit (CPU)** is where data is processed. When buying a computer, you will see the **clock speed** of the CPU, measured in megahertz (MHz) or gigahertz (GHz), advertised. Usually, the faster the clock speed, the faster the data is processed.

The computer's **memory** is associated with the CPU. There are two types of memory chip: **read-only memory (ROM)** and **random-access memory (RAM)**. Data in ROM cannot be changed and is permanently stored even when the computer is turned off. ROM is used to store the programs needed to start up the computer when it is switched on. RAM is used to store software and data while they are in use.

Figure 12.3 USB port

The processor is housed in a box. **Peripheral** devices (this is the general term to describe input, output and storage devices) can be attached to the processor via **ports** – the sockets that can be seen on the side of the computer. Modern computers have **Universal Serial Bus (USB)** ports into which a variety of devices can be plugged. With USB, a new device can be added to a computer without an adapter card having to be added. The computer does not even have to be switched off.

Processors come in a variety of forms. The most common are desktop personal computers, laptop computers and personal digital assistants (PDA).

▶ Software

Software is the name given to computer programs. Programs are made up of thousands of instructions that actually make the computer do the processing that is required. Without software, computer hardware is no use.

▶ Information

Information is data that has been processed or converted to give it meaning to a human being. An itemised receipt for a customer produced at a supermarket checkout is an example of information. See Chapter 13 for a detailed description of information.

▶ Procedures

ICT systems usually involve a number of procedures that have to be undertaken to make sure that the system can run smoothly.

An ICT system that uses a large database will need established backup procedures to ensure that if the system fails for any reason the data could be restored. A procedure to archive data when it is no longer current is also needed. **Archiving** is the removing of data that is not currently needed by the system to an offline file. For example, in the attendance system described in Example 1, the data may only be kept online for a term. Attendance data for the previous term would not be deleted but archived in case it were to be needed.

When a user wishes to transfer photo images from a digital camera to their computer they need to carry out the procedure of connecting the two devices with a cable. If they wish to print out paper copies of certain photos they must load the printer with the appropriate paper.

Specific ICT tasks generate procedures that need to be carried out. In the attendance system example, when a teacher enters data about whether pupils are present or absent, she is carrying out a procedure.

▶ People

ICT systems also involve people. Most data is entered by human beings – either through use of a keyboard or by another device such as a touch screen or a bar code scanner. The information output by the ICT system is read and used by human beings

A user needs to carry out general procedures such as backing up data, changing and disposing of printer cartridges and logging on to the computer at the start of a session.

Consider the student attendance system described in Example 1. The information produced by the system is only of use if the teachers have entered data to indicate which students are present in the classroom. It is also a task for someone to distribute the printed reports produced by the system to the appropriate people.

case study 2
▶ booking a seat at the cinema

The cinema in Wigtown has five studios. Bookings can be made by telephone, over the Internet or in person. Customers are assigned numbered seats. For telephone bookings, the booking clerk asks the customer for the name, date and time of the film showing that they wish to attend and the number of seats to be booked. If there are the required number of seats available, the customer is asked where in the studio they would like to be seated. Once the seats have been chosen, the booking clerk selects them from the screen and the booking is made.

Payment is made using a credit card. The booking clerk asks the customer for card details and enters them into the ICT system. When authorisation is obtained, the booking is confirmed.

The customer may request the tickets to be sent by post, in which case the clerk prints out the tickets using a special printer loaded with blank tickets and enters the name and address of the customer into the system so that an address label can be produced. The tickets are then put into an envelope, the address label is added and the letter is put in a pile to be franked and posted at the end of the day.

When a booking is made over the Internet, the customer selects the film, date and time for themselves as they are presented with a series of menus. They are presented with a diagram showing the available seats and can highlight their choice.

A number of procedures can be identified in the scenario described in this case study. Most of the procedures are carried out by the booking clerk. For example, when tickets are sent out by post, the booking clerk carries out a procedure to print out the tickets, enter the address details, print out the address label and put the tickets in an envelope. Another procedure, of franking the envelopes and posting them is carried out by the clerk or someone else.

1. What is meant by "franking a letter"?

2. Identify other procedures carried out by the booking clerk.

Activity 2

For each of the scenarios below identify the procedures and people involved in the ICT system:

■ The cost of products purchased by a customer at a supermarket is calculated (see case study 1).

■ A purchase is made in a shop using a chip and pin card (see case study 1).

■ A digital audio player is loaded with tracks from a number of CDs.

■ An organisation of 5000 employees produces monthly payslips.

SUMMARY

▶ **ICT is the use of technology for the input, storage and processing of data and the output of information.**

▶ **A system is an activity or set of activities that involves input, processing and output.**

▶ **A system that uses any form of digital technology is an ICT system.**

▶ **The output from an ICT system may go directly to a human being or into another ICT system.**

▶ **An ICT system consists of the following components: data; hardware; software; information; procedures and people.**

▶ **People are a necessary part of ICT systems. They carry out procedures such as checking and entering data or distributing printed outputs.**

▶ **Hardware is the name given to the physical components of the computer or communication system.**

▶ **Software is the name given to computer programs.**

▶ **Information is data that has been processed or converted to give it meaning to a human being.**

Questions

1. Name and describe three components of an ICT system giving an example of each. (9)

2. Explain what is meant by an ICT system. (2)

3. An ICT stock control system for a shoe shop keeps a record of every item held in stock together with the item's current stock level. The stock level of a product is modified whenever new stock is delivered or a pair of the shoes is sold.
   a) Identify two items of data that are input into the system. (2)
   b) Describe two processes that are carried out by the system. (4)
   c) State one example of information that might be produced by the system. (1)
   d) Explain who would use this information. (2)

4. ICT systems use both hardware and software. Explain, using examples, what is meant by hardware and software. (6)

5. A customer goes to a fast food restaurant and orders a meal for his family. He pays using his "chip and pin" credit card. The restaurant has an ICT payment system.
   a) Identify the two people who play a role in this ICT system. (2)
   b) For each of the people identified in a), describe a procedure that they would carry out. (4)

What do we mean by data? ◀

The term "data" means recorded facts – usually a series of values produced as a result of an event or transaction.

For example, if I buy an item in a supermarket a lot of data in the form of facts is collected such as:

- my loyalty card number
- the identity numbers for each item bought (often called the bar code number)
- the weight of apples to be purchased
- the number of the credit card used to pay for the goods.

All this data has been generated by an **event** – me buying some items in a supermarket.

If I pay a cheque into my bank account, this is an event that collects a record of the **transaction**. The record might contain details such as:

- my bank account number and sort code
- the bank account number and sort code for the cheque paid in
- the amount of the cheque.

Other examples of data include:

- Account numbers, such as 0244 78200 04191 1 or 07379082
- Postcodes, such as DE13 0LL and W1A 1AA
- Bar codes, such as 50 00231 036422 and 31 05634 412412
- Numbers, letters, names and dates.

The meaning of the data may not be obvious; on their own, items of data may not be much use to a person. However, data is very useful when it is processed to create information.

Figure 13.1 Examples of data

How data can arise: direct and indirect data capture ◀

Data capture means the collection of data to enter into a computer. Data can be input into a computer in a variety of ways, depending on the source:

- keyboard
- speech recognition, using a microphone
- webcam
- touch screen
- selection from list using a mouse
- sensors
- bar-code reader
- transfer of pre-captured data from an external device e.g. a digital camera.

Data can be captured directly or indirectly.

Direct data capture is the collection of data for a particular purpose. Examples of direct data capture are:

- reading bar codes at a supermarket till so that the product can be identified
- account details being read directly from the chip embedded in a credit card
- an MICR device automatically reading the numbers on the bottom of a cheque
- data from an automatic weather station being downloaded into a computer.

Indirect data capture is the collection of data as a by-product from another purpose. Examples of indirect data capture are:

- using data from reading bar codes at a supermarket till to work out stock levels
- using the records of transactions generated at a store when a customer uses a loyalty card to build up a profile of the buying habits of a customer. The store could sell this profile of a customer to another company, enabling them to target mail at the customer for products that the customer is most likely to buy.

Encoding data ◄

A modern computer must store data of many types: for example, the data may be in the form of text, pictures, sound or numbers. When data is input into an ICT system it has to be converted from its current form into a form that can be processed digitally. This means that it has to be represented by a binary code made up of binary digits (bits), which can be written as either 0 or 1.

The term "encoding data" means putting the data into an appropriate binary code. The following are all examples of files stored in the computer in binary code:

- text, such as DOC files
- digital pictures, such as bitmaps, JPG or GIF files
- digital videos, such as MPG or AVI files
- sound files, such as WAV or MP3 files.

Different binary coding systems are used to represent different types of data. The smallest unit of storage is called a **bit**. A bit can be in one of two states: one state is represented by a 0, the other state by a 1. By building up combinations of bits, different codes can be stored. Two bits can store any one of four different codes: 00, 01, 10 or 11. Three bits can store one of eight codes: 000 100 001 101 010 110 011 111. Four bits can store 16 codes, 5 bits 32 codes, 6 bits 64 codes, and so on.

As well as data, program instructions are also stored in binary code of 0s and 1s. The same bit pattern of, say, 16 bits could represent two text characters, an integer (a whole number), part of a graphical image or a program instruction.

The program that is running interprets the bit pattern appropriately.

Activity 1

Figure 13.2 The code that represents an image shown in a text editor

1. Create a very small digital image (you could crop an existing image) and save it (for example, in JPG format).
2. Load up text-editing software, such as Notepad.
3. Open the image file in Notepad.
4. You can see the machine-readable code for the image. It is not in binary but each character of the code represents a longer binary code.

Why do you think the output is as it is?

You can try this with sound files, such as WAV, as well.

▶ Text

Data made up of words or symbols is stored in text or character form. The word processor is the most common software package that processes text data.

The following phrase is an example of text and is made up of 36 characters – 34 letters and 2 spaces:

Information Communication Technology

Each character (including a space) that can be used is assigned a unique binary code. Standard codes have been agreed so that data can be transferred and correctly interpreted between two ICT systems.

The most widely used character coding system is **ASCII** (American Standard Code for Information Interchange). A character coded in ASCII is made up of 8 bits. The 8th bit acts as a check bit to help ensure that any corruption of data is detected. The other seven bits can produce 128 unique codes, each a different combination of 0s and 1s. Thus 128 different characters can be represented.

Activity 2

In ASCII, the term "ICT" is coded:

01001001 01000011 01010100

1. Use a search engine, such as Google, to search for "ASCII codes in binary" to find a table of the codes for characters. Using this, write your name in ASCII code. (Remember that the space character also has a code.)
2. Do you speak a language other than English? Does it use different symbols? Can you find the binary code for appropriate symbols in Unicode? Try http://www.unicode.org.
3. How many bits are allocated for a character in Unicode?
4. How many different codes can be stored in Unicode?
5. Why is Unicode sometimes used rather than ASCII?

► Images

Increasingly, computers are used to manipulate, store and display non-textual images. For example, a photograph can be downloaded from a camera, stored and used on a web page. Computer games can contain complex graphical images. Indeed, the most commonly used interfaces on a personal computer are made up of graphical images in the form of icons. These pictures, or graphics, also have to be stored in binary coded form.

There are two main ways in which images are stored: either as bitmapped or vector graphics.

Bitmapped graphics

A bitmap is the binary stored data representing an image. A picture is broken up into thousands of tiny squares called **pixels**. The number of pixels stored for a given area determines the **resolution** of the image. The more pixels that are used per square unit, the greater the resolution of the

Figure 13.3 Greyscale image of lion

image. The greater the resolution of an image the better the image looks. The greater the resolution of the image, the more memory is required to store its bitmap.

Each pixel is allocated a number of bits in the bitmap to represent its colour. The more bits allocated to each pixel the greater the choice of possible colours, but the amount of memory required to store an image also increases.

If only two colours, black and white, are used then just one bit is needed to represent each pixel. A 0 can be used to represent white and a 1 to represent black. If four colours are to be represented then two bits are needed for each pixel and the coding could be, for example:

00 – white, 01 – red, 10 – green, 11 – black

Figure 13.3 shows a greyscale image built up in pixels.

As processors have become faster, both the main memory (RAM) and the backing storage capacity have increased hugely. Modern computers are able to store and process complex images of high resolution that are made up of many colours.

Bitmapped graphics can be created by using a drawing package where individual pixels can be set or lines "drawn". This is achieved using some kind of pointing device to modify an image displayed on a screen. Alternatively, an image can be input using a scanner or a digital camera. A software package can then be used to modify the image. For detailed changes, the setting of individual pixels can be modified.

There are a number of standard formats that are used for storing graphical data. These are necessary, in the same way that ASCII is necessary for text storage, to allow graphical data to be transferred between different packages. An image developed in a painting package may then be used in a desktop-publishing (DTP) package. One common format is Tagged Image File Format (TIFF).

Compressed bitmap files

Data compression techniques are used to minimise the amount of storage space needed for graphical images. The Joint Photographic Experts Group (JPEG) has defined standards for graphical image compression. JPEG is now a commonly used format.

Problems associated with bitmapped graphics

■ Bitmapped graphics can be difficult to edit. For example, if a line needs to be redrawn, the pixels in the deleted line have all to be changed to the background colour and the pixels that make up the new line have to be changed too.

- Image quality can be lost if enlargement takes place as the size of pixels is increased and the resolution of the image (the number of pixels per unit area) is reduced.
- Distortion can occur if the image is transferred to a computer whose screen has a different resolution as pixels can be elongated in one direction.
- A large storage space is required to store the attributes of every pixel.

Activity 3

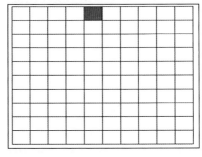

Figure 13.4 Grid of pixels

Draw the image that is represented by the following binary codes. The image is in four colours and two bits are used to store the code for one pixel. The coding is:

00 – white, 01 – red, 10 – green, 11 – black

The codes for the pixels are stored line by line and are printed in blocks of 8 bits for easier reading.

00000000 01000000 00000000 00010101 00000000 00000101 10010100
00000000 00010101 00000000 00000000 01000000 00001111 00000010
00001111 11111100 00100011 11110011 11110010 11111100 00001111
11101111 00000000 00001011 00000000

This starts 00 (white) 00 (white) 00 (white) 00 (white) 01 (red) 00 (white) 00 (white) 00 (white), and so on.

Vector graphics

For applications such as computer-aided design (CAD), where high precision is required, bitmapped graphics are not appropriate. With vector graphics, the image is stored in terms of geometric data. For example, a circle is defined by its centre, its radius and its colour.

Vector graphics enable the user to manipulate objects as entire units. For example, to change the length of a line or enlarge a circle, the user simply has to select the chosen object on the screen and then stretch or drag the image as required. The bitmapped graphic requires individual dots in the line or circle to be repainted. Using vector graphics, objects are described mathematically so they can be layered, rotated and magnified relatively easily (see Figure 13.5).

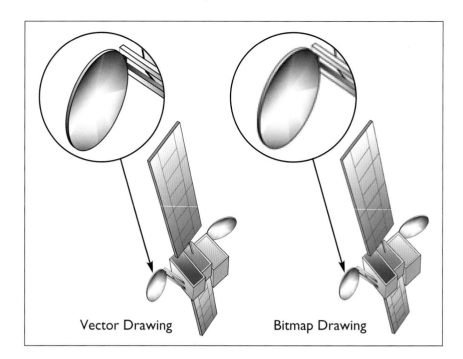

Figure 13.5 Effect of enlargement on a vector and a bitmap graphic

Vector Drawing   Bitmap Drawing

case study 1
▶ Trinity House Lighthouse Service

Trinity House provides nearly 600 aids to navigation sites, such as storm-lashed lighthouses and buoys. Project teams consisting of specialist engineers are responsible for projects from initial design through to completion.

A CAD program is used to generate detail and assemble drawings that are used in the manufacture, construction and installation of navigational equipment. The CAD program uses vector graphics. This enables the engineer to represent the various components and services as objects that can be copied or adapted to serve different applications. Layering can be used to differentiate services such as water supplies, electrical cabling and control systems.

Input is normally via a digitiser and tablet with the primary output device usually being a pen plotter or A3 laser printer.

1. Explain why a bitmapped graphics package is not appropriate for use by the Trinity House Lighthouse Service. Give at least three reasons.

2. Explain why a digitiser and tablet is used for input rather than a mouse.

Moving pictures

Animations consist of a number of images or frames stored together and displayed one after the other. The more frames and the smaller the change between frames, the more realistic is the effect of the animation. An animation stored as GIF files can take up considerable storage space.

The MPEG file format uses a method of compression for video information in a similar way to that used in JPEG files for single images, the aim being to eliminate repetition between frames. MPEG files also allow a soundtrack. In spite of the size reduction resulting from compression, even a short piece lasting only a few minutes will have hundreds or thousands of frames and so the file size is likely to be large.

▶ Numbers

Computers store numbers, as all other data, in binary coded form. There are three main ways of coding the numbers that we use.

Integers are whole numbers, such as 7, 24567800, –56 or 0. When integers are stored in a computer, the number of bits assigned to the code determines the range of numbers that can be stored. One byte (8 bits) can store positive numbers in the range 0 to 255, while two bytes (16 bits) can store positive numbers in the range 0 to 65535.

The coding can be designed to store negative as well as positive integers. Integer arithmetic provides fast and accurate results; problems only occur if a calculation results in an integer that is too large to be stored in the number of bits assigned to the code.

A **real number** is a number that can have a fractional part. Unlike integers, real numbers can rarely be stored exactly in the bits assigned to store the number. Try dividing 100 by 3 (by hand) and you will find you can never write down all the digits after the decimal point. You could write down 33.3 or 33.33 or 33.333333333. Whatever you write will not be exact; the representation of real numbers in a computer always involves some loss of accuracy. The more bits allocated to store a real number, the greater the range of numbers that can be stored, in the same way as with integers, but the accuracy of the representation of the number also increases. Performing calculations on real numbers is a more complex operation than performing calculations with integers, and therefore is slower.

In systems where fractional values are needed but where accuracy is very important, such as when data is representing money, then a third form of coding can be used. There are a number of applications where numbers that represent currency values are stored in a special format.

▶ Boolean values

A **Boolean** value is one that can take one of only two values – one representing true (or yes) and the other false (or no).

▶ Sound

Sound travels in waves and is therefore analogue in form. To be stored in a computer, the analogue signal must be converted into digital form. The wave that is input through a microphone is **sampled** at regular intervals by an analogue-to-digital converter. This device measures the height of the wave at the time of sampling and stores this as a binary code. The number of times that the wave is sampled for a given time period is known as the **sampling rate**. The more frequent the sampling, the more accurate the representation of the sound. The amount of storage space required increases as the sampling rate of the sound increases.

Typically, music stored on an audio CD has 44,100 samples per second, each sample using 16 bits with two channels (for stereo sound). This means that a CD stores about 10 megabytes of data per minute of music. A five-minute song therefore requires 50 MB of data.

Data compression techniques are used to reduce the amount of storage space required. MP3 is a standard coding system using compression techniques that stores the sound files in a smaller space. MP3 can compress a song by a factor of about 10 while keeping close to CD quality. The 50-MB sound file is reduced to about 5 MB when stored in MP3 format.

When the sound is output, the digital representation is converted back to analogue form and the signal is output through a speaker. How closely the sound resembles the original wave depends upon the sampling frequency.

In the same way that there are a number of standard formats for storing text and graphics, so too there are standards for sound storage. WAV is the standard audio format for Windows and AIFF for the Macintosh. Both platforms can also play and save sounds in the AU and SND audio formats as well as the MIDI format which is specifically for music. MP3 format, described above, is used to download and store music files from the Internet.

Why things go wrong ◀

If data is entered incorrectly, whether accidentally or deliberately, then the information output will be incorrect. Information is only as accurate as the data that is entered. If the data source is wrong, the information output will be wrong. This is sometimes referred to as garbage in, garbage out (**GIGO**).

Stories abound of things going wrong with computers. A warehouse production system at Rootes Group in the 1960s didn't know the difference between feet and inches – some components turned out 12 times bigger than they should have been.

NASA made a similar mistake when trying to send a rocket to Mars. Some measurements were in inches and some in centimetres. Let's just say that the rocket didn't land as expected.

▶ Coding data

Data has to be processed to produce information. In some circumstances, a code is used to represent the data to allow processing to be more effective and to produce more useful information for the user.

Common examples of the use of coding include:

■ Gender is usually stored as M or F instead of Male or Female.
■ Banks use branch sort codes such as 60-18-46 instead of the name of the branch.
■ Dates of birth, such as 6 February 1986, are coded as 06 02 1986 instead of as 6th February 1986.
■ Airline baggage handlers use codes for destinations e.g. LHR means London Heathrow and FRA means Frankfurt.
■ Postcodes identify a small number of buildings. For example, SO9 identifies an area of Southampton and SO9 5NH is the university. If the code is written on an envelope, it can be converted into a series of dots for automatic sorting.

Codes are used because:

■ They are often easy to remember, such as LHR for London Heathrow.
■ They are usually short and quicker to enter; fewer errors are likely to be made.
■ They take up less storage space on disk.
■ They ensure that the data stored is consistent. For example, it would be difficult to search a file for people with a given date of birth if some were stored as 25th May 1988 and others as May 25 1988.
■ It is easier to check that data is valid if the range of options is limited.

example	It is possible to access a computer program on the Internet that outputs the full address when you enter a postcode and house number. Entering the postcode:
	■ is quicker to type in as it is much shorter than the full address
	■ allows an immediate check to be made to see if the postcode really exists
	■ avoids spelling the name of street incorrectly.

Data is entered into the computer and stored in code. However if the data item is to become information it must be decoded before it is output. So at a supermarket till, the bar codes of the items are entered into the computer as data but information – the total cost for all the purchases – is displayed on the till.

▶ Loss of precision due to coding

Market research questionnaires often ask the subject to tick an age range (see Figure 13.6). The answer is stored as a code, for example, *c* for anyone who is 25 to 34. This coding is easy to use but leads to a loss of precision.

- People who are 25 are bracketed with people who are 34.
- If you had to work out the average age you wouldn't be able to do so.

The fewer the categories, the greater the loss of precision.

Questionnaire

Are you: Male ☐   Female ☐

Are you: a. Under 18 ☐   b. 18–24 ☐   c. 25–34 ☐   d. 35–44 ☐

e. 45–54 ☐   f. 55–64 ☐   g. Over 65 ☐

Figure 13.6 Age-range question on questionnaire

▶ Worked exam question

Discuss the input, processing and output required for the production of wage slips for a small bakery.

(6)

▶ **EXAMINER'S GUIDANCE**

There are six marks available. It is sensible to assume that two marks can be gained for discussing input, two for processing and two for output. To gain two marks for discussing input it is necessary to say more than just what data is to be input. Something more is needed – perhaps something about how the data can be entered. Remember, it is a small bakery.

Describing the processing means going further than just saying, "The process is to work out the wages."

Again, your answer concerning output needs to give more than the information that is produced – perhaps you could describe the form that it takes.

▶ **SAMPLE ANSWER**

One example of data input is the number of hours worked by each employee in a week. (1) As the bakery is small, this data is typed in at a keyboard. (1)

One process needed to fulfil the task would be to multiply the hours worked by the rate per hour for each employee. (1) This will calculate their gross pay. (1)

One example of information output is the pay for each employee. (1) This will be printed on a payslip. (1)

What is information? ◀

Information is data that has been processed and given a context, which makes it understandable to the user.

For example, when bar codes are scanned at a till, they are simply data being entered into a computer. The computer program looks up the name of the item and its price and calculates the total amount payable which is then displayed on the till. This is information because the data has been processed and a context added so that it is meaningful.

Both of the examples shown in Figure 13.7 are meaningful to the user.

Information is said to be of good quality if it is accurate, up to date and relevant for a particular use.

Figure 13.7 Examples of information

▶ Accurate

How accurate it needs to be will depend upon the use being made of the information. For example, the information in a bank statement must be exactly right to the nearest penny otherwise the account holder could make inappropriate decisions over spending and would be fully justified in complaining to the bank. When reporting overall A-level pass rates at a school, a figure to the nearest 1% would usually be sufficiently accurate.

The accuracy of information is very important. For example, inaccurate stock figures may cause a store manager to re-order the wrong amounts. This could result in items not being available for customers or too much stock being held which would take up extra space and tie up capital.

▶ Up to date

If information is out of date, then wrong decisions can be made. It is very important that all reports produced include a date so that the person receiving it knows how old it is. In certain circumstances, a time is also required. This enables the reader of the report to know exactly when it was produced. Information changes over time and without a date the reader might make a wrong decision, unaware that the information is out of date.

If a list of names and addresses used for mail shots is five years out of date, many letters will be sent to a wrong address as the person named may have moved or even died during that time. It would be a waste of time and money for a company to send out letters to these addresses. The older the mailing list, the fewer the "hits", and consequently the fewer responses there is likely to be.

In order to determine the number of AS ICT classes to run in a college, the administrator would need to know the number of students wishing to follow the course this year. Last year's figures might be quite different; if they were used instead of this year's numbers, then too few classes might be set up.

A sales manager for a large company selling office furniture throughout the UK will need to decide which salesmen have been doing badly and need extra support and which products to choose for extra production or promotion. Using information based on an earlier month's figures rather than the most recent could result in incorrect decisions that could lead to reduced sales and profit.

When an employee of a cinema is taking a booking for the evening performance of a film, he needs to know which seats are available at the time of booking, not the ones that were available at the start of the day.

Information has to be up to date to be useful. Keeping the information up to date affects the costs of producing the information. The costs arise from the need to:

- collect up-to-date data
- enter the data into the system
- delete out-of-date data.

Traditionally many computer systems operated in **batch** mode, in which data was collected over a period of time, a day, a week or even a month before processing took place. The information then produced was only as up to date as the most recent processing run. **Transaction-processing** systems update with new data as it arises, processing each change as it occurs. Information produced from a transaction-processing system is always as up to date as possible, but such a system is more costly to run as faster processing, more sophisticated hardware and faster communications might be required.

An organisation needs to have systems in place to ensure that changes in data can be collected. This can prove to be very complicated and time consuming as changes can include altered marital status and perhaps a change in surname, a change in telephone number as well as an address change. Very often when a patient attends a doctor's or dentist's clinic, the receptionist will quickly check basic data with them to ensure that it is up to date.

It is very important that a school or college has up-to-date information about its students. Current contact addresses and telephone numbers are essential. A system would be in place whereby students could fill in a form whenever any changes occur. However, such a method would not be foolproof. As a way of catching missed changes, every student might be given a printout of their personal details and qualifications every year, or even every term, for checking. Any changes would be added and the student would be expected to sign the form. The data held electronically would then be updated. Such a process would cost money. Each student's details would need to be printed out and distributed; missing students would need to be chased up; every sheet would need to be checked for changes and the changes entered. The data could be kept even more up to date if the printouts were issued and checked every week but the cost in both time and money would make this unproductive for the few changes that would be highlighted each week.

Many organisations try to gather address changes and other details from customers by including a form with every invoice that allows the customer to enter changes easily. It is then a simple matter for this changed data to be entered onto the database.

Name and address lists used for targeted mailing soon get out of date. Systems need to be put in place that enable entries to be deleted when no longer current and new names to be added. Organisations which keep a database of customers will probably record a date when the customer last made contact in some way, by making a purchase or an enquiry. In this way, people who have made no contact for a set period of time can be deleted from the database.

▶ Relevant

Information that is essential in one situation may have no use in another. For example, information intended for a branch manager of a supermarket, showing checkout till usage to allow him to allocate staff over a weekly period, would not be of use to the regional manager wanting to see the efficiency of all branches.

A figure showing the percentage pass rate in A-level ICT for all pupils in a school would be useful for teachers when planning and reviewing their work. However, a detailed list showing each student's mark in every module would be needed when advising individual students whether or not to re-sit modules.

Activity 4

For each of the following situations, state why the information might lead to an inappropriate decision being made. Describe the possible consequences of the decision.

1. The purchasing manager of a mail-order company is looking at a list of stock levels for his products before placing orders prior to the Christmas rush. The list was produced in June.

2. Two trained nursery teachers are considering setting up a new nursery school in the local area to start in September 2008. They have information from the local council of the number of children under the age of eleven in the local area which was produced in June 2004.

3. A bakery wishes to determine ways of increasing profits. They want to know whether they could sell more loaves and which are the most popular. The manager is given the number of loaves of each type, baked each day for the last 12 months.

4. The manager of a holiday company has to pre-book places in hotels for the 2008 season. He has the results of a market-research survey of holiday preferences from members of the public that was carried out in 2002.

▶ Data means raw facts and figures; records of transactions.

▶ Data can arise directly or indirectly.

▶ Direct data capture is the collection of data for a particular purpose.

▶ Indirect data capture is the collection of data as a by-product from another process.

▶ The term encoding data means putting the data into an appropriate binary code. Different binary codes are used to represent different types of data.

▶ Data made up of words or symbols is stored in text or character form. The text is usually stored in ASCII.

▶ Images can be stored as either bitmapped or vector graphics.

▶ A bitmap is the binary stored data representing an image.

▶ Vector graphics stores the image in terms of geometric data.

▶ Coding systems exist for different types of numbers: integers (whole numbers), real numbers (numbers with a fractional part) and currency (pounds and pence in the UK).

▶ Sound has to be sampled at regular intervals. The number of times that the wave is sampled for a given time period is known as the sampling rate.

▶ Data frequently needs to be coded when collected to enable effective processing.

▶ Information means data which has been processed to give it a context which gives it a meaning.

▶ A poor data source leads to poor information.

▶ Good information should be accurate, up to date and relevant for its particular use.

▶ Ensuring that information is up to date can be time consuming and costly.

Questions ◀

1. Explain the difference between data and information. (4)

2. Explain the difference between direct and indirect data. (4)

3. A checkout operator in a supermarket scans the bar codes of items being purchased by customers. The scanner is linked to an electronic point-of-sale (EPOS) system. The software that is used contains functions to look up the prices and descriptions of the products that are scanned in order to produce an itemised receipt for the customer. The software also produces a daily sales summary report for the store manager. State a data item that is entered into the EPOS system and describe two items of information that are produced. (5)

4. Data input to an ICT system can take many forms such as pictures, sounds, numbers and letters. In all cases, the data has to be encoded. Explain why data needs to be encoded. (2)

5. The expression "garbage in, garbage out" (GIGO) is often used in connection with information-processing systems. Explain what is meant by this expression. (4)

6. A report has been produced by an ICT system for the sales manager of a company. He then complains that he does not know when the report was produced or how up to date the contents of the report are.
   a) Explain why up-to-date information is important to the sales manager and what could be done to ensure that he knows when the information was produced. (4)
   b) Discuss other factors that affect the quality of the information produced for the report. (8)

7. A supermarket stock-control computer system updates its stock levels every evening based on that day's sales. Describe the possible consequences of the supermarket using out-of-date data. (4)

8. The owners of a holiday company are considering organising some special deals for 2009. They use data obtained from the customers who took one of their holidays during 2005 to decide what to offer. One of the owners, James, says that the information produced is not suitable for making the decisions.
Explain what makes James unsure of the usefulness of the information and discuss the implications to the holiday company of making decisions based on it. (6)

People and ICT systems
AQA Unit 2 Section 3 (part 1)

◄

Characteristics of users ◄

There is no such person as a "standard ICT user". Different users have differing requirements which depend on a number of factors:

- Experience
- Physical characteristics
- Environment of use
- Task to be undertaken
- Age

► **Experience**

Some users will use a particular ICT system on a regular basis. They will become familiar with using the system and will want to be able to carry out tasks as quickly as possible. They are likely to be irritated by carrying out operations that slow them down, such as having to make selections from numerous sub-menus before reaching the function they require, waiting while an introductory screen with music is displayed when the system is loaded or having to repeatedly carry out the same actions without the facility to save entries for later use.

Other users will use an ICT system infrequently. For example, they may order goods from a particular web site just once or twice. They will not build up any expertise in using the system and would become frustrated if it was not made very clear how the system should be used.

Users who rarely use any ICT systems generally find it much harder to use a new system than those who have worked extensively with a range of other systems.

Activity 1

1. Identify three ICT systems that you use on a regular basis (see Chapter 12 for what is meant by an ICT system). For each, list features that are appropriate for you as a regular user and things that irritate you.
2. Identify three ICT systems that you have used only a few times. For each, describe how easy the system was to use and identify the features that made it so. Suggest any improvements that could be made to help an inexperienced user.

► Physical characteristics

Users have different physical characteristics. Some users have poor eye sight, others may lack manual dexterity and be unable to use a standard keyboard.

ICT offers many opportunities for people with disabilities, particularly those who have difficulty communicating. There are various computer adaptations available for people who cannot use a mouse or keyboard or who cannot read from a normal monitor very well.

Someone who can operate a pointing device, such as a mouse, but not a standard keyboard can use an on-screen keyboard. This provides point-and-click access to standard keyboard letters, whole words and communication phrases.

A person who is unable to operate a keyboard or a mouse may use a computer system with speech recognition.

To assist those with poor eyesight, output can be to a large screen, spoken by a speech synthesiser, or to a special printer in the form of Braille.

Software and web sites should be designed to make them accessible to as many users as possible.

► Environment of use

While many computer systems are used by people sitting at a desk, this is not true of all ICT systems. Consider a system that allows a train passenger to buy their ticket at a machine. They require a system that is quick to use, is robust and has clear instructions. An ICT system using a touch screen would be appropriate.

When a person is on the move they may wish to access their emails using a mobile phone. As the display screen is small, the interface needs to be very simple and clear to use.

► Task to be undertaken

If an application requires large amounts of text to be entered then the most appropriate form of interface will probably involve a keyboard. However, human–computer communication is not just about entering data at the keyboard and reading text on the screen. A voice recognition system may be more appropriate.

Adventure games have video-quality graphics and CD quality sound. A keyboard or mouse would not be quick or precise enough as an input device – a joystick is needed.

► Age

An ICT system for use in a primary school will need screen designs to be very simple with visual clues as the children may not be able to read very well. A concept keyboard with pictures overlaid on a pressure-sensitive pad may be more suitable than the standard keyboard.

<table>
<tr>
<td>

case study 1

▶ **Microsoft Vista accessibility**

</td>
<td>

The operating system Microsoft Vista offers a range of features that can make the computer accessible to a wider number of users. The software provides a feature where a user can adjust accessibility settings, such as magnification or the use of special keys, as well as manage special accessibility programs that are included:

- The user can enlarge part of the screen image in a separate window using a magnification program called Magnifier.

- Narrator converts on-screen text into speech using a natural sounding voice.

- An on-screen keyboard can be displayed in one of several layouts. The keyboard can be configured to a suitable font.

- A speech recognition program is also included.

- A user can replace system sounds with visual images such as a flash on the screen.

1. A user with poor eyesight might wish to use the magnification program. What other methods of input and output could be suitable for a partially sighted person?

2. Explain the circumstances when a user might decide to use speech recognition software and describe the uses to which this software could be put.

3. Find out more about the accessibility features of Microsoft Vista.

4. Search the Web for information and produce a presentation on one or both of the following:

   a) ICT systems for users with visual impairment

   b) ICT systems for users with physical disabilities.

</td>
</tr>
</table>

Human–computer interface (HCI) ◀

Most ICT systems involve human interaction at some point, for example:

- A person ordering goods online needs to select items and fill in a form – probably using a keyboard and a mouse.
- A passenger may purchase a ticket for a train journey at a ticket machine using a touch screen to make selections from a set of menus.

- A garden designer may create or modify an idea for a garden layout using a CAD package on a computer with a large screen and a graphics tablet.
- A sales assistant in a supermarket may process a customer's purchases using an EPOS (electronic point-of-sale) till with a flatbed, bar-code scanner.

The **human–computer interface (HCI)** is the point of interaction between people and computer systems. The HCI should be designed to make it as easy as possible for humans to communicate with the computer. In particular, creating an appropriate HCI requires:

- the choice of appropriate hardware devices (both input and output devices)
- designing the "look and feel" of the software including screen layout design.

▶ Designing the HCI

The design of how humans and computer systems interact is crucial to the successful use of a computer system. The choice of HCI will depend on the application and the needs of the user.

The earliest computers had interfaces that were not user-friendly and could only be used by people with extensive technical knowledge. As the capabilities of computers have increased, HCIs have improved.

Graphics are now used extensively in HCIs. As the memory capacity and the processing speeds of computers have increased, the use of graphics has become more widespread. **Icons** – tiny pictures designed to convey an easily understood meaning – are fast and easy to interpret and are not language specific.

Sound can be a feature of an HCI. Audible error messages can alert a user, for example, to pressing the CAPS LOCK key. Printer drivers can include sounds so that users get audible messages such as "Please load paper in the cut sheet feeder" or "There is a paper jam in the printer".

An HCI should be easy to use, appropriate for the users, safe and robust. The interface should be user-friendly. This means that the software should be easy to use and new features easily learned. It should have a consistent "look and feel": wherever users are in the software they should be presented with things in the same way. For example, the date might always appear in the top right-hand corner or messages pointing out when data has been incorrectly entered should be given in a standard way. If common functions are accessed in a similar way across a range of software packages, a user can transfer already acquired skills when using new software for the first time.

<Ten rules for good interface design>

1. All system interfaces should be consistent and should look, act and feel the same throughout. Keep the layout clear. Too much on one screen will make it cluttered and confusing.

2. Put features such as headings, error messages and requests for user responses in a consistent location.

3. Make good use of colour and contrast. Do not, for example, use red on a green background: people with colour blindness cannot differentiate between them.

4. Make sure your text is not too big or too small. Stick to one clear font. If possible, offer the user the option to change the size of the text.

5. Start at the top left of the page and work down.

6. Don't use jargon, particularly with inexperienced users.

7. Whenever the user is deleting work or closing a program, confirm all requests for action from the user; for example, "Are you sure? Y/N".

8. Make sure that the choices in a menu are easily distinguished from each other. Leave plenty of blank space between them.

9. Keep the level of complexity appropriate to the experience of the expected user. If the level is too high, the user will become confused. If it is too low, the user will be bored and make errors due to lack of attention.

10. Use the same name throughout for the same thing. For example, stick to the word "save" and don't use "keep", "store" or "write".

case study 2
▶ a web interface

Many websites provide customers with services including the opportunity to purchase goods. Most users only use a particular site on an occasional basis and so the interface needs to be easy to use.

The web interface must be designed to be simple to use because if it is not the customer could become frustrated and abandon using the site all together, taking their custom elsewhere. Creating an appropriate interface is an essential element in supporting good customer management.

The web site http://www.greatgifts.org/ is designed for occasional users who wish to purchase unusual gifts. The gifts can be chosen from a wide range and can be selected by category or by price range.

Figure 14.1 shows how the use of drop-down boxes and menu choices makes the system easy to use as all the possible options are presented to the user in a familiar format.

Figure 14.1 Getting started to choose a gift

In Figure 14.2, the details of specific gifts are displayed. Once the customer has selected a gift by clicking on the "select gift" button, a wizard takes them through the remaining stages of the purchase.

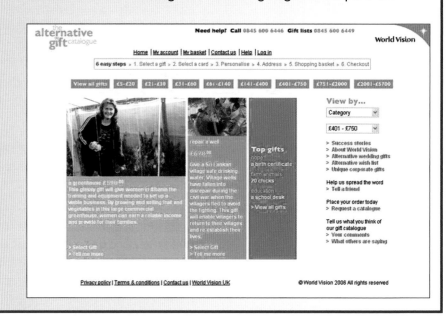

Figure 14.2 Selecting a gift

case study 2 (contd)

They need to select an appropriate card, choosing the image for the front of the card as well as the message to go inside (see Figure 14.3).

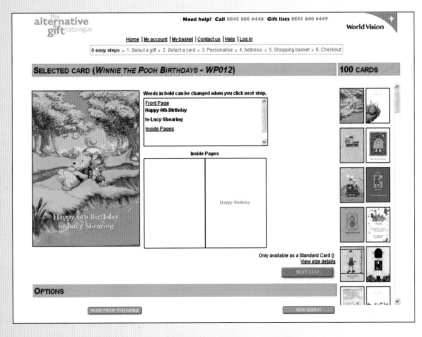

Figure 14.3 Choosing a card

The system requires the customer to set up an account so that the necessary data can be collected for the order. Each time a gift and card are chosen, they are added to the customer's shopping basket. When all the gifts have been chosen, the customer clicks on the checkout button and is then asked for details of how the goods are to be paid for.

When the details of the transaction are completed the customer is asked to verify that he wishes to make the order. This provides the customer with confidence that they will not be able to change an order if they press the wrong key by mistake.

The site offers the user a chance to read their privacy policy. Many users are wary of ordering goods online as they fear that their personal data could be misused.

Access the web site and locate the Privacy Policy page.

1. What promises does World Vision make concerning personal data?

2. How and why is personal data stored?

3. What steps does World Vision take to ensure that customers' personal data is kept safe and secure?

4. Explore the help facilities offered. How useful would they be to a customer?

5. Discuss how well each of the ten rules for good interface design has been met by this website.

6. Explain what features of the ICT system make it easy for occasional users and thus promote good customer management.

► Worked exam question

Explain two factors that you think should be considered when designing a web interface for use by customers that would support good customer management.

(4)

AQA Specimen

You can refer to case study 2 for help with this question. Many of the customers using a web site will not have used the site before, or will be infrequent users. The customers are likely to have very varied skill levels.

Many customers have worries about the use of such e-commerce sites. The HCI should help to address some of these worries.

► SAMPLE ANSWER   Customers will have different levels of ICT skills. (1) A simple menu system would give specific options to enable those with little skill to use the system. Appropriate online help should be provided. (1)

Many customers are afraid of safety when using an online system for ordering and paying for goods. (1) The HCI needs to offer assurance that the site is secure. (1)

► Online help facilities

Nearly all recently produced software includes **on-screen help** for the user to enable them to use unfamiliar features. Even an experienced user needs help when using a feature of the package for the first time. The help screens should be written in clear English, avoiding jargon whenever possible.

Adequate and consistent help should be given to novices. Many software applications include **wizards** which provide a novice with prompts that take them through a particular task. The use of a wizard can make complex procedures within the system easier for the user. A wizard breaks down a task into manageable steps, each step having a clear explanation of what data the user needs to enter.

Such help is particularly appropriate when a user wants to use a feature of the system that they have not used before, or needs to be reminded how to use an infrequently used function. A user may wish to check how certain data that has to be entered needs to be formatted. The wizard in Figure 14.4 is from Microsoft Word and allows a user to set up all the formatting and features for a letter.

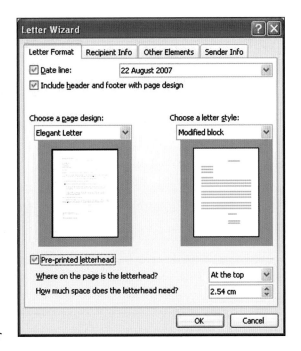

Figure 14.4 A Microsoft Word wizard to produce a customised letter

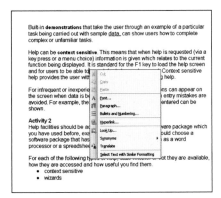

Figure 14.5 A pop-up menu obtained by right-clicking the mouse

Some software also offers **tips** or **assistants** that are displayed when the user is carrying out a particular task. Such tips can provide the user with an alternative way of carrying out the task. Built-in **demonstrations** take the user through an example of a particular task being carried out with sample data. They can show users how to complete complex or unfamiliar tasks.

Help can be **context sensitive**. This means that when help is requested (via a key press or a menu choice), information is given which relates to the current function being displayed. It is standard for the F1 key to load the help screen and for users to be able to search for help on a key word. Context-sensitive help provides the user with a consistent method of gaining help.

For infrequent or inexperienced users, detailed explanations can appear on the screen when data is being entered to ensure that data entry mistakes are avoided. For example, the format in which a date is to be entered can be shown.

Activity 2

Help facilities should be accessible to users. Using a software package which you have used before, examine the help facilities. You should choose a software package that has a large range of features, such as a word processor or a spreadsheet.

1. For each of the following types of help, state whether or not they are available, how they are accessed and how useful you find them:
   - context-sensitive help
   - wizards
   - demonstrations
   - tips
   - other (please specify what they are).
2. Comment on the appropriateness of the language used.
3. Use the help facilities to learn how to use a function that you have not used before.

case study 3
▶ online class registration

A school makes use of an online registration system that allows teachers to take class registers in the classroom on a computer. Each teacher enters a unique identification code supported with a password. The attendance data is stored centrally and links to the school's student database.

The ICT support department manages the system, carrying out tasks such as:

- adding new teachers and pupils

- entering details of classes and the pupils in those classes

- making modifications.

Class tutors and senior managers receive regular reports on attendance. These include weekly attendance lists for each tutor group or form as well as statistical summaries. Pupils and their parents are able to access a summary of their own attendance online through the school's intranet.

Each category of user needs the facility to gain help when necessary so that they can make good use of the features of the software.

Teachers have access to on-screen help that relates to the stages they go through to enter the attendance for a class. They quickly become familiar with the common day-to-day functions that they need. There are however some other functions, such as registering an absent colleague's class, that teachers only use occasionally. On-screen instructions that remind them of the data that has to be entered help them to use the software in this situation. Every teacher has a printed instruction sheet highlighting the main procedures of the system for use after long school holidays!

The members of the ICT support department have technical manuals that include details of all functions in the system. They set up and

case study 3 (contd)

maintain the appropriate data on an annual basis. From time to time they need to phone the help desk of the software house when a problem occurs that they cannot solve for themselves. An online user group has been set up, where those supplying technical support to different schools and colleges share their problems through a bulletin board.

The tutors and managers are provided with a printed guide showing the reports that the system can produce and how to interpret each one. The guide includes sample reports.

The pupil or his parent accessing through the Internet is given simple online instructions explaining how to access and interpret the information relating to their own attendance. A step-by-step guide is provided, helping the user to complete all the actions required, showing exactly what data needs to be entered in every field. There are explicit instructions displayed on the screen to make data entry as simple as possible.

■ Copy and complete the table below. Include further ideas of your own as well as those given in the case study.

User	Use of software	Typical help needs	Appropriate sources of help
Teacher	Entering class attendance data		
Form tutor			
ICT support			
Pupil			

case study 4
▶ theatre booking system

A software house has produced a seat-booking software package for sale to theatres and cinemas. The system allows front of house staff to receive seat bookings over the telephone and face to face before a performance. The tickets produced list the title of the film or play, the date and time of performance, the seat number and the price. Customers can also book online by accessing a web site.

The theatre or cinema manager has to enter details of the forthcoming programme into the system. The manager can also produce statistical reports showing such things as the sales of seats, the popularity of different shows and the cumulative takings.

1. Draw up and complete a table similar to the one for online class registration.

2. Some customers choose to order face to face or by telephone because they worry that the web site may not be secure, which could result in a third party getting hold of their credit card details. Describe other reasons which might deter a customer from ordering goods online.

3. What are the advantages of ordering goods online?

User support

◄

Most ICT system providers offer some form of support for users in case they have difficulty in installing or using the system. Support is sometimes provided free under the product warranty or an entitlement to help can be bought for a fixed period of time. Large organisations have an ICT support team whose members provide in-house system support for users.

Users may also need access to help on a regular basis. The ICT system may offer many features, not all of which the user makes use of on a regular basis. There needs to be a suitable way for the user to find out what he or she needs to know at the appropriate time.

► Telephone help desks

Many ICT system suppliers offer telephone support for immediate help and advice. This provides someone with technical skills to guide the customer.

This help may be available during business hours; for some widely used, general-purpose software, the help could be available 24 hours a day. The user phones the help desk (or call centre) when they have a problem. Help desk operators are trouble-shooters who provide technical assistance, support, and advice to customers and end-users. They are experts in the ICT system and are likely to have a computer on the desk in front of them, which they will use to try to replicate and solve the users' problems.

When a help desk is provided for a widely used system, it is likely that a high number of requests for help will be received. The help desk provider will need to establish procedures for logging and tracking the requests for help and the advice given. This is to ensure that all requests are dealt with in a fair and timely manner to maintain customer satisfaction.

► Email support

If the problem is not time critical, then email could be used as an alternative to the telephone. This has the advantage of smoothing out the demand, so that the operator can answer queries in order throughout the day. A priority system could be used which would ensure that critical enquiries were answered first. Operators will be able to spend all their time finding solutions to problems without being interrupted by a ringing telephone.

From the user's point of view, the use of email avoids wasted time on the telephone. However, instant answers to simple problems are not possible. Email lacks the opportunity for human interaction offered by a telephone conversation.

Many suppliers now offer instant messaging support, as one operator can deal with several sets of instant messages at the same time.

Online chat support is also widely offered as an interactive method of support that is an accessible alternative to telephone support. Online chat can provide a user with real-time access to help and advice of other users or employees of the software producers.

▶ User guides

An ICT system provider usually creates written instructions for using a package. These are typically provided free with the system. This user guide may come in hard copy form as a book or in a soft copy that can be stored on the user's hard disk so it can be accessed whenever required. The user guide will describe how to install and use the system. For complex systems, there can be a number of different books: perhaps one aimed at a first time user, in the form of a tutorial, and another involving a complete description of all functions to serve as a reference document. These guides allow users to work at their own pace with the instructions beside them so they can find out how to use the system functions for themselves.

In an organisation some people, perhaps senior managers, may not be directly using a system themselves. However, the system may produce reports that will provide them with information that they will need to make appropriate decisions. They will need to know what reports the system is capable of producing and how to interpret them. These managers would require a written manual that provides instructions for using the reports, together with samples of the report so that they can determine which reports will aid them in their decision making.

▶ Online support

A popular way of making support available to the user is to use the Internet. Help facilities can be stored on the Web. Information can be kept very up to date. Users can access **patches** to update software or fix errors. Such a site would offer the facility to email a package expert for advice on a specific problem.

▶ Support options for industry standard packages

Many industry-standard packages, such as Microsoft Office or Adobe Photoshop, have a very large user base. Between them, the users need to know all aspects of the software's functionality, although each individual user is likely to use only part of the range of options a software package offers. Support is provided in a variety of ways.

Help is now usually available over the Internet where up-to-date advice on common problems can be stored.

Software houses provide a considerable amount of information on packages via **specialist bulletin boards** on the Internet, often in the form of **frequently asked questions (FAQs)**. Users can search through questions that other people have asked and are likely to find a solution that resolves their own current problem. The software publishers provide websites with information about the packages and

case study 5
► CensorNet support options

Adelix CensorNet software provides internet safety and management facilities to users. It allows organisations to keep their employees safe online. It defends against all threats spread by email.

Figure 14.6 shows three options for support of the software that Adelix provides.

Figure 14.6 Adelix support options

■ Discuss the factors that a manager would consider when deciding which of the three options to choose when purchasing and installing the software.

■ Explore the online chat facilities available for a software package that you use, such as Microsoft Office. These can usually be accessed from the "Help" menu.

■ Investigate software that is available for organisations to set up their own online chat facilities. Try http://www.click4assistance.co.uk.

email access to experts. It is often helpful for the user to print out a version of the advice provided for use at a later time when the same information is needed. Such facilities are particularly useful for more able users who can help themselves by reading the information provided.

For complex systems, user groups are set up, where users can get together to share problems and ideas. Such groups can meet physically or, more often these days, virtually via bulletin boards on the Internet.

Books about using popular software are produced independently by publishers and sold in most bookshops or over the Internet. These provide the user with the opportunity for self-help. For the most popular packages there are a very large number of titles available. A user needs to ensure that any book purchased is designed for a user of their skill level as books vary from those suitable for absolute beginners to those designed for the most advanced user. Publishers of widely used software may send **newsletters** to all registered users including support articles on tips, solutions to common problems and advanced functions. These newsletters provide a forum for users to share ideas and problems.

Many of these facilities are of most use to the experienced user who, given access to the appropriate information, can work out the solution to a problem for themselves. Very often the help that a less experienced user needs is closer to hand. It is possible that a colleague or friend who is familiar with the software can help with their problem.

Activity 3

1. Explore the Help facilities that are available for a spreadsheet package such as Microsoft Excel. List the ways of obtaining help within the package.
2. Describe any online help facilities that are available.
3. Explore http://office.microsoft.com/ and list the ways of receiving help with the use of Excel.
4. Explain how you would go about obtaining a newsletter with information about Microsoft Excel.
5. Find the titles of three books on Excel: one introductory, one for users with some experience and one for advanced users. The following web site may be of use: http://www.compman.co.uk/. Summarise the contents of each book.
6. Explore the following site: http://www.compinfo-center.com/pcsoft/spreadsheets.htm#top. List any support facilities available to users of Microsoft Excel.
7. Can you find any further useful sites for Microsoft Excel users on the Web?

SUMMARY

Different users have differing requirements dependent upon:

▶ **Experience**

▶ **Physical characteristics**

▶ **Environment of use**

▶ **Task to be undertaken**

▶ **Age.**

The HCI must be designed specifically for a given environment: different situations and users at different levels need very different interfaces.

System providers offer a range of support options to customers:

▶ **Help desks**

▶ **Email support**

▶ **User manuals**

▶ **On-screen help.**

Much support is expensive to provide. However, an ICT system needs support if it is to maintain credibility with the public.

Industry-standard software packages have a very large number of users. Other methods of obtaining help are available to users of these packages:

▶ **Newsletters**

▶ **Bulletin boards on the Internet**

▶ **Frequently asked questions (FAQs)**

▶ **User groups**

▶ **Books.**

Questions

1. After logging in to a network, the desktop screen appears as shown in Figure 14.7. Describe three ways in which this screen provides an effective human–computer interface. (6)

June 2004 ICT2

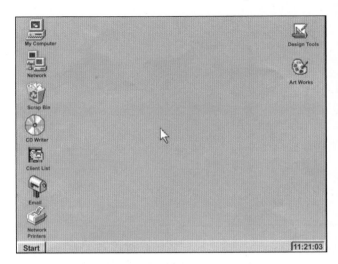

Figure 14.7 A network desktop

2. Describe how the interface between a computer and a human can be designed to provide an effective dialogue. (9)

3. A new employee of a company is unfamiliar with the word-processing software used by the company.
 a) Describe the help facilities that she would expect to find as part of the software. (6)
 b) Discuss two further sources of help that might be available to her. (4)

4. A mail-order company receives orders from its customers, handwritten on pre-printed forms. These are then used by clerks to enter the data into a computerised system. A sample blank customer order entry screen is shown in Figure 14.8. With reference to Figure 14.8, describe three features of the input screen that provide an effective human–computer interface. (6)

June 2002 ICT2

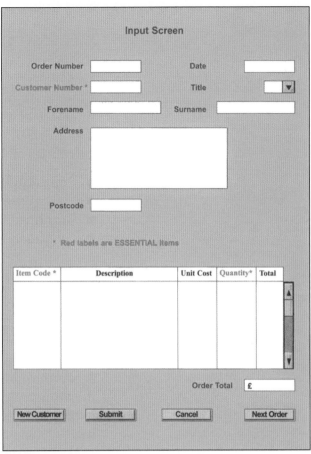

Figure 14.8 Customer order entry screen

5. A user purchases a copy of a widely used spreadsheet package from a computer store. Describe three methods of gaining support from sources other than the software's manufacturer. (6)

6. ICT systems are designed for and used by people. Discuss how the characteristics of users may vary. (12)

15

User interfaces

AQA Unit 2 Section 3 (part 2)

◀

Chapter 14 discussed the importance of designing a suitable interface of an application so that it provides for effective communication for users. There are several standard categories of user interface:

- Command-line interface
- Menu-based interface
- Graphical user interface
- On-screen forms

Each of these is described in this chapter together with its benefits and limitations.

Command-line interface (CLI) ◀

A command-line interface is an interface where the user types in commands for the computer to interpret and carry out. For example, to run a word-processing program, the user may have to type in WORD.

As the command is typed in, it appears on the screen. The user has to know the commands and there are no clues to help guess them.

The MS-DOS operating system uses a command line interface. The screen is usually blank and C:> (called the C prompt) appears on the left of the screen. This means that the computer is looking at drive C, the internal hard drive. Any file references typed in refer to that drive.

Some operating systems, such as Unix and Linux, use both a command-line interface and a graphical user interface. Figure 15.1 shows a typical command line in the Unix operating system. The **ls** command lists the files in the current, or specified, directory. It has many options; **-l** produces a long (detailed) listing.

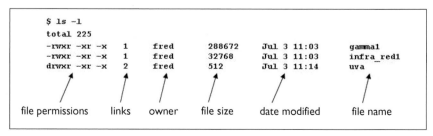

```
$ ls -l
total 225
-rwxr -xr -x   1     fred     288672    Jul 3 11:03    gamma1
-rwxr -xr -x   1     fred     32768     Jul 3 11:03    infra_red1
drwxr -xr -x   2     fred     512       Jul 3 11:14    uva
```

file permissions   links   owner   file size   date modified   file name

Figure 15.1 Unix directory listing command

Many commands are complex. There may be additional parameters, usually extra letters at the end of the command, that modify its meaning.

Little computer memory is required for a command-line interface. Complex commands can be entered quickly in one line. Precise sequences of instructions can be entered allowing complex tasks to be performed. The first computers, with tiny memory and limited processing power, used command-line interfaces.

The user has to learn all these commands. As a result command-line interfaces are normally only used by experienced and expert users. For someone with little experience, a command-line interface can be very frustrating and frequent reference to manuals may be needed.

case study 1
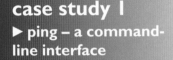
▶ ping – a command-line interface

When network developers are installing a new network, they use a utility program called "ping" to test that one computer on the network is connected to another.

Ping sends a small packet of information to a specified computer, which then sends a reply packet in return. From this reply, the ping program can check whether you can reach the other computer and how long it takes to get the reply.

Ping uses a command-line interface; users can do a variety of different tests by typing in different commands.

You can use the MS-DOS prompt on a PC to experiment with ping:

■ In **Windows Vista**, click on **Start**. In the **Search** box, type **cmd**. Press **Enter**.

■ In **Windows XP**, click on **Start > Run**. Type in **cmd** and press **Enter**.

■ In **Windows 98**, click on **Start > Run**. Type in **command** and press **Enter**.

The MS-DOS prompt window opens.

These commands may be disabled on your school or college computer.

Figure 15.2 MS-DOS prompt window

case study I (contd)

To test a connection to another workstation on a local area network, type in "ping" followed by the IP address of the other workstation e.g. ping 192.168.0.2. Figure 15.3 shows that a connection has been found.

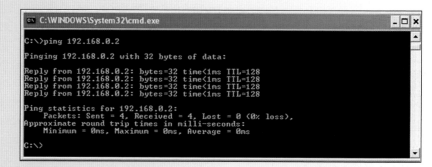

Figure 15.3 Output from the "ping" command

If you are not connected to a local area network but are connected to the Internet, try ping followed by a known URL, e.g. ping www.google.co.uk (see Figure 15.4).

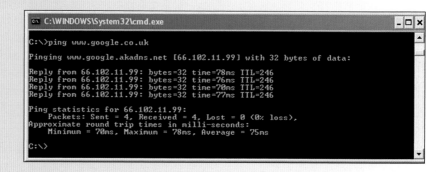

Figure 15.4 Output from the "ping www.google.co.uk" command

Menu-based interface

A menu-based user interface is a form of interface that displays a set of choices on the screen so that the user can make a selection from the range of choices offered. They can operate in a number of ways:

- The selections may be numbered and the option selected by pressing a number key.
- The user may scroll through the selections using the cursor keys or a mouse until the required option is highlighted and then press enter.
- If a touch screen is used, the user touches the screen at the place where the chosen option appears.

Choosing an option may lead to a sub-menu being displayed. A balance needs to be made between the number of options on a screen at one time and the number of levels of sub-menu

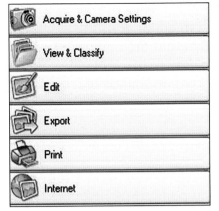

Figure 15.5 The main menu from ZoomBrowser Ex, a photographic utility

required. The more levels needed, the longer the system takes to operate and the user is more likely to get irritated.

However, too many options in one menu makes it harder for the user to make the correct selection. Care must be taken to make sure that sub-menu options logically link together. If they do not, a new user may become confused as it can be hard to know which sub-menu to pick when searching for a particular option.

You have probably seen menu-based user interfaces in the following applications:

- a mobile phone
- an MP3 player, such as an iPod
- a digital camera
- a bank cash machine.

A well-designed, menu-based interface should have a consistent layout. It should use the same prompt for the same operation in all menus and the prompt should be in the same position on the screen.

---

case study 2
▶ mobile phones and HCI

Mobile phones use menu-based user interfaces. There are a variety of different interface techniques to make phones easy to use. These include:

- A hierarchy of menu and sub-menus
- An address book so that you can select commonly used numbers from an alphabetical list of names
- In the address book, pressing a letter takes you to all the names beginning with that letter
- Voice recognition to select a name from the list
- Predictive text to make it quicker to enter text messages.

1. Draw a diagram to show the sub-menus on your mobile phone.

2. Would you make any changes to the structure of the menu to make it more easy for you to use?

3. Design a menu structure for an MP3 player or a digital camera.

▼

A menu-based interface can only deal with situations where the user's requirements are known in advance as the user is limited to the predetermined choices that are built into the menus. A menu-based interface is easy to use and appropriate for relatively inexperienced and occasional users of a system.

Graphical user interface   ◀

A graphical user interface (GUI) is a form of interface that uses high-resolution graphics, icons and pointers to make the operation of the computer as user-friendly as possible. The aim of the interface is to make its use intuitive for a user; this is often achieved by building on real-life ideas such as a window or a desktop, where different documents or tasks are displayed on the screen in the form of icons in a similar way to work on a real desk. The use of sound and video are also made use of in many GUIs. Options can be chosen easily with a pointing device – usually a mouse.

A GUI is sometimes called a WIMP environment as it uses:

- Windows
- Icons
- Menus
- Pointers.

GUIs were first developed for the Apple Macintosh but soon afterwards Microsoft Windows was developed as a GUI for the PC. GUIs tend to need a lot of memory and disk space and take time to load because of the large number of graphical images used. However, today's computers are more than powerful enough to cope with a GUI. The main operating systems used nowadays offer the user a graphical user interface.

case study 3
▶ railway ticket machines

Automatic ticket machines are common at many railway and underground stations (see Figure 15.6). These machines use a GUI to allow customers to purchase a ticket quickly and easily without having to queue in a ticket office. Customers use a touch screen to choose their destination from a list and the type of ticket, such as single or return. Payment is by credit or debit card: the ticket machine can automatically read the card details from the magnetic strip on the back.

A touch screen is used because:

■ It is more durable than other pointing devices, such as a mouse.

■ It is easy to operate, even for the inexperienced user.

The instructions for use are displayed on the screen and are very simple to follow. The user has few decisions to make. The HCI is robust and very easy to use.

Figure 15.6 A railway ticket machine uses a GUI

1. List five other situations when it would be appropriate to use a touch screen, justifying your choice in each case.

2. Design the screen layout for one of the situations you have listed. Take into account the rules of good interface design listed in Chapter 14.

▶ Windows

A window is a rectangular division of the screen that holds the activity of a program. There can be several windows on the screen at the same time. The user can switch between windows and change the size and shape of the windows. The active window – the one in which the user is currently working on – appears at the front (see Figure 15.7).

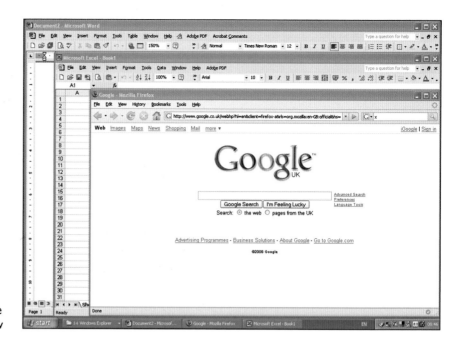

Figure 15.7 Three windows on one screen – Google is the active window

A **dialogue box** (spelt dialog box in the USA) is a window that appears on the screen when information is wanted from the user.

For example, a **wizard** can be used in Microsoft Access to create a report (see Figure 15.8). Several dialogue boxes appear on the screen, one after the other, asking the user to choose settings so that the report appears as the user wants.

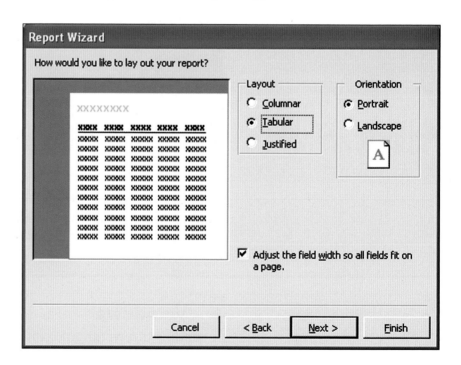

Figure 15.8 A dialogue box in the Microsoft Access Report Wizard

► Icons

An icon is a small picture on the screen. Clicking on the icon performs an action such as saving a file. The same action can be performed using the menu system but the icon is used as a shortcut. The action performed when you click on an icon should be easily recognisable from the image.

Today, icons are used in nearly all software. Icons used in different programs are often very similar, as most software uses icons to open files, save files, etc.

Icons can be grouped together in toolbars on the screen. In many application packages, icons and toolbars can be customised to suit the user. Icons can be added or removed. The image on an icon can be edited.

Software such as Microsoft Word offers the facility to display icons in a large format (see Figure 15.9). This is useful for people with a visual impairment and those who find it difficult to click on a small icon.

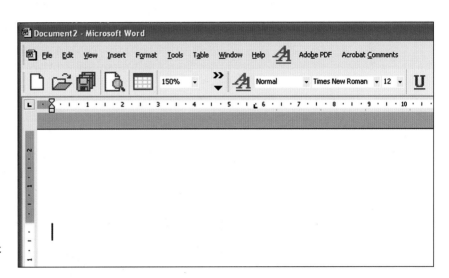

Figure 15.9 Large icons in Microsoft Word

► Menus

A **pull-down menu** (see Figure 15.10) is a menu that expands downward when selected with a mouse or other pointer. The user then scrolls through the options and clicks a second time to make the selection. Windows software typically has a menu bar of pull-down menus across the top of the screen.

Figure 15.10 A pull-down menu

A **pop-up menu** is a similar menu that expands upward when clicked on with the pointing device. In Microsoft Excel and Microsoft Word, the Drawing toolbar typically appears at the bottom of the screen. When you click on an item on this toolbar, a pop-up menu appears.

In most Windows software, if you click the right mouse button a shortcut menu appears. If you right click near the top of the screen, a pull-down menu appears. If you click near the bottom of the screen, a pop-up menu is displayed.

In many application packages, menus can be customised to suit the user. Options can be removed, new options added, whole menus removed or new menus added.

▶ Pointers

A **mouse** is a very common pointing device. It can move a cursor around the screen to be used for selecting a choice from a menu or pointing to any place on the screen. A **tracker ball** may be used in a similar way. However, there are other pointing devices associated with GUIs. A laptop computer usually has a built-in **touchpad**; a PDA has a **stylus**. If a computer is to be used by members of the public, a mouse may not be robust enough. In this case, a **touch screen** is often used.

On-screen forms ◀

On-screen forms (see Figure 15.11) are widely used to enter data into a computer system. Forms are used to capture a standard set of data items, for example, when ordering goods or filling in a questionnaire.

On-screen forms reduce the chances of an error because the user is prompted to fill in each field in turn. Forms can be designed so that if an essential field has been left blank, the form is not accepted until this field has been filled in.

It is a good idea for an on-screen form to mimic the style of a paper form, so that it is intuitive for an inexperienced user to fill in. For example, on-screen forms should be designed to be filled in from left to right and from top to bottom, as paper forms are. On-screen forms should also enable the user to go back to make changes.

On-screen forms can include **check boxes** for yes/no fields and can allow the user to choose from a menu or list, where appropriate. This has several advantages: it restricts the user to allowable options, reduces the possibility of error and is usually faster. Choices can be made from a menu using a **drop-down** (or combo) box or by using option buttons.

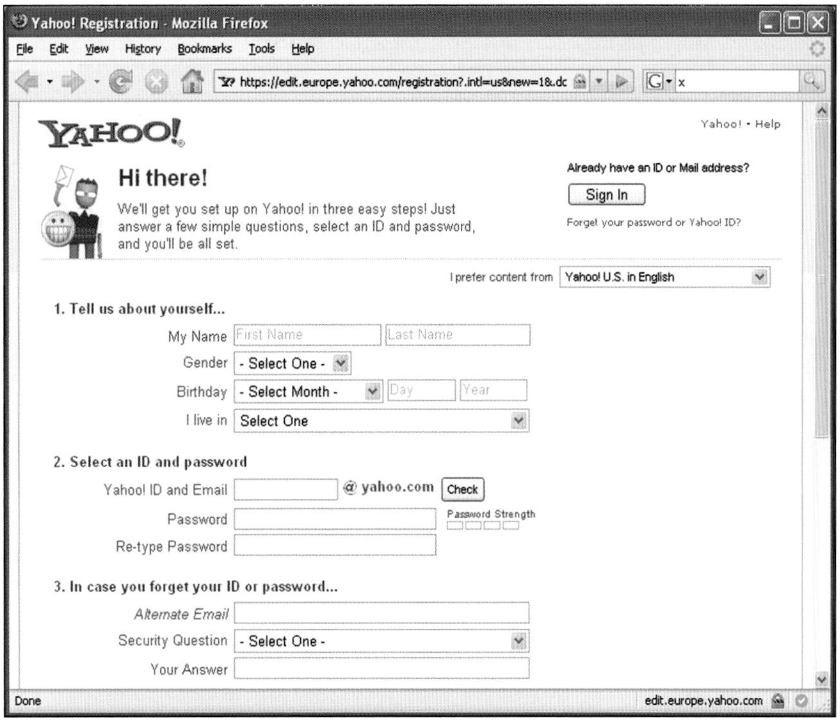

Figure 15.11 An on-screen form at Yahoo Mail

Another way of speeding up data entry is to use **default** settings – the most likely entry is provided and it can be accepted by pressing a key to move the cursor to the next field on the form.

If there are a lot of questions, more than one screen per form is needed and the form should be split up into logical divisions. For example, a paper booking form for a holiday might require information on customer, holiday destination, mode of travel and car hire all on the same page. This would not fit onto one screen. The form could be split up to have one screen for each of the sections mentioned.

Activity 1 – looking at on-screen forms

Search the Internet to find three sites each of which has an online form for a customer to fill in.

1. For each form, state whether or not you feel that the form is well designed and suitable for the user. Describe features of the design to support your argument.
2. Design an on-screen order form for a company selling T-shirts online.

Table 15.1 Benefits and limitations of types of interface

	Benefits	Limitations
Command-line interface	• Little computer memory required • Wide variety of commands • Commands entered quickly on one line	• Commands are complex • The user has to learn all the commands
Menu-based interface	• Easy to use • Uses little computer memory • Can include shortcuts • Suitable for devices with small screens, such as mobile phones	• Only suitable when there is a choice of actions • There may be several levels of sub-menu
Graphical user interface	• Easy to use and intuitive • Flexible – many uses • Uses graphics to show meaning • Can use a variety of pointing devices • Can be customised to suit the user • Can include on-line help	• Needs a lot of memory • Needs more disk space than command-line or menu-based interfaces • Takes time to load
Form-based interface	• Similar in style to paper forms • Easy and intuitive to use • Can use a variety of data-entry techniques such as drop-down boxes • Can be automatically validated	• Needs a lot of memory • Needs more disk space than command-line or menu-based interfaces • Takes time to load

case study 4
▶ travel agency

A travel agency has an ICT system that provides information to help customers choose and book their holidays. Different types of user use the system in different ways.

Customers visiting a branch of the travel agency can interrogate a local offline system to find details of holidays. The use of a touch screen might be appropriate so that the user can simply select from a number of choices on the screen to obtain the information they require. The user is not required to enter any data other than their choices. The layout of the screen should be simple and uncluttered to attract the user. The text should be large and good use could be made of colour. There could be a built-in printing device so that the customer can print out the details of any holiday that meets their requirements.

Travel agents based in the branches use the system to make bookings on behalf of clients. Although the agents are regular users, using this system is only part of their job, so it needs to be straightforward to use. Choices need to be made concerning such factors as country, destination resort and travel dates. Further textual data about the customer and members of the party are also required.

ICT specialist staff at the head office set up the system and maintain the accuracy of the database. These users are expert and spend the majority of their working time using the system. They are likely to be using the system in a variety of ways and to be very knowledgeable about its workings.

case study 4 (contd)	Consider the interface required for each user and complete the grid below:				
User	Most appropriate type of interface	Input devices required (with reasons)	Output devices required (with reasons)	Discuss special features required of interface	
Customer				Need clear, uncluttered screen	
Branch agent					
Head office specialist					

Natural-language interface ◀

In the real world, humans communicate in natural languages, such as English. A natural-language computer interface, where humans and computers could communicate in normal English would be enormously advantageous. A natural-language interface is appropriate for use by a search engine.

Natural language is extremely flexible and powerful. However, to interpret natural language easily, a computer needs it to be precise with a limited number of commands and strict syntax rules.

English is not like this. It is not precise and has a huge vocabulary. New words appear and the meanings of other words change over time. Sentences can be ambiguous. It is only when the context and tone are taken into account that the meaning is clear.

It is not yet possible to develop systems that accept a full natural language but some systems have been developed using a limited subset of a natural language by restricting instructions to a limited number of commands.

Commands are usually typed in at a keyboard but, if such a system is combined with speech recognition, there is potential for humans and computers to have a limited dialogue.

Pure natural-language interfaces may still be the stuff of science fiction but there are many research projects exploring the possibilities of talking to computers. The Spoken Language Systems (SLS) Group at the Massachusetts Institute of Technology (MIT) Laboratory of Computer Science has been trying to get computers to communicate in the same way as humans since 1989. Progress has been made, as can be seen from the excerpt from a conversation between a user and JUPITER, an SLS-based weather forecasting system.

example
▶ natural-language interface

> User: Yes I would like the weather forecast for London, England, please.
> JUPITER: In London in England Wednesday partly cloudy skies with periods of sunshine. High 82 and low 63. Is there something else?
> User: What is that in degrees Celsius, please?
> JUPITER: In London in England Wednesday partly cloudy skies with periods of sunshine. High 28 and low 17. What else?

Check their progress at http://www.sls.lcs.mit.edu/sls/.

The search engine Ask Jeeves (http://ask.com/) uses a system called Natural Language Processing (NLP) that allows the user to ask questions in natural language. It builds up a bank of questions and uses these to help interpret new ones. With NLP, Jeeves appears able to understand the context of the question and so can offer answers and suggestions in the same human terms in which we all communicate.

In Microsoft Office software, such as Microsoft Word, a help dialogue box appears when you press F1. You can type in a natural-language question to get help.

One of the problems in creating a natural-language interface is the fact that natural languages, such as English, can be ambiguous. For example, the word "lead" has several meanings. It can mean the leash for a dog. It can mean the person in front in a race. Pronounced differently, it can mean the writing part of a pencil. The written sentence "I want the lead" could mean:

■ I want the leash for my dog.
■ I want the lead to put in my pencil.
■ I want to be in front.

The word "by" is also interpreted in different ways in different contexts:

■ The lost children were found by the searchers (who)
■ The lost children were found by the mountain (where)
■ The lost children were found by nightfall (when)

The structure of a sentence can also be ambiguous. If you were to say "My car needs oiling badly" would you really want someone to make a bad job of oiling your car? What do you make of the sentence "Fruit flies like a banana"?

The human–computer interface (HCI) considers how people communicate with computer systems. The choice of HCI depends on the application and the needs of the user.

Common interfaces are:

▶ command-line interface (CLI)

▶ menu-based interface

▶ graphical user interface (GUI)

▶ form-based interface

▶ natural-language interface

A command-line interface (CLI) is a user interface in which the user responds to a screen prompt by typing in a command. The system displays a response to the screen and the user then enters another command. This process is repeated.

A menu-based user interface displays a set of options allowing the user to make a selection from the range of options offered.

A graphical user interface (GUI) is a form of user interface that does not rely on text. It requires high-resolution graphics. GUIs usually have the following features:

▶ Windows

▶ Icons

▶ Pull-down and pop-up menus

▶ Pointers

A GUI is easy and intuitive to use but demands fast processing speeds and a large computer memory.

On-screen forms are widely used to enter into a computer system. Forms are used to capture a standard set of data items, for example, when ordering goods or filling in a questionnaire.

It is very difficult to enter instructions into a computer system in a 'natural language' such as English. This is because computer instructions need to be very precise and English has a large vocabulary which can be ambiguous.

Benefits of a natural-language interface:

▶ It is the natural language of humans, who can express themselves freely without constraint

▶ There is no need for special training.

▶ It is extremely flexible.

Limitations of a natural-language interface:

▶ Natural language is ambiguous and imprecise.

▶ Natural language is always changing.

▶ The same word can have different meanings.

Questions ◀

1. Most modern PCs make use of a graphical user interface (GUI). Describe four characteristics of a GUI. (8)

2. Explain what is meant by the term "command-line interface". (2)

3. An Internet search engine is said to have a natural-language interface. Discuss the benefits and limitations of using a natural-language interface. (6)

4. WIMP (windows, icons, menus and pointers) interfaces are commonly in use by software packages on personal computers. Name two other types of interface and discuss the benefits and limitations of each. (12)

5. The owners of a small business that sells badges have decided to start selling their products online. They will collect order data from their customers using an on-screen form at their website. Discuss the factors that they should consider in designing the form to enable customers to fill in the data easily. (8)

Working in ICT
AQA Unit 2 Section 3 (part 3)

Perhaps the majority of jobs now require the use of ICT in some way. For most people ICT is a tool that helps them carry out their job. An ICT professional is a person who works in the development, maintenance or support of ICT systems. Many ICT professionals have formal qualifications in ICT.

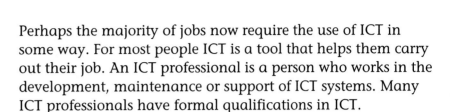

PRODUCT DATA ANALYST
Intelligent? Efficient team player with attention to detail? Strong data analysis skills? Excel a must and Access DB a plus. Leading e-marketing company based in Leeds. Small team environment. Fun work culture. Excellent package.
Send CV to ICT@ICT.uk or fax 020 56789

WEB DEVELOPER
XHTML : PHP
CSS: MYSQL
Central London marketing agency working on creative projects. Salary on application.

Senior IT Support Person
Package 30K Plus Exciting role providing Pre and Post sales support to our customers. Visit www.itsupport.com to see our software solutions.

Based in Cumbria

We are also recruiting Sales People with experience in Technical Sales OTE 35K Plus
No agencies

IT Support
£24–28k + Bens
IT Security company is looking for a Support Engineer to provide internal/customer support. Solid expertise in Microsoft networking. Exp in following a plus: fw's, anti-virus, IDS, TCP/IP. Training provided.
Please send CV to jobs@jobs.com

Technical Consultants

You will have a background of application design, specification and integration.

You will be able to research and analyse technical requirements at all levels and match these to the needs and constraints of the business.

You will have the capacity to produce creative and cost effective solutions within a complex customer environment. You will be confident with a consultative approach and be an effective communicator with business and technical people.

Figure 16.1 ICT job adverts

Activity 1

Use newspapers, trade magazines such as Computer Weekly, or the Web to search for job advertisements in the following fields:

- Systems analyst
- Programmer
- Database developer
- Web developer
- Help-desk administrator

For each job category explain what the job involves.

case study 1
▶ Mary – computer help-desk operator

Mary is a member of a team of people who work on an ICT help desk in a large national museum. The help desk is a department within the museum that responds to users' technical questions about any ICT system installed within the museum. These questions can vary from a request to find out how to use a function that the user has not encountered before to troubleshooting an unexpected error.

Mary works closely with her colleagues. Between them, the team members have to ensure that all requests for help are dealt with quickly and efficiently so that the users can get on with their work. If Mary is unsure of the necessary response to a user, she will consult another member of the team.

Mary can provide help in a number of ways. She can talk to the person using a telephone or, more frequently, using email or instant messaging. In some cases, she can take remote control of the user's computer to work through the problem.

1. Explain the term "instant messaging".

2. Why is instant messaging an appropriate means of communication in this instance?

3. Explain why effective team working is necessary for the help desk to run efficiently. Think of some ideas of your own as well as the factors discussed in the case study.

4. Describe any help-desk facilities available at your school or college.

5. Find out the five most frequent problems dealt with by the school or college help desk.

There is a range of jobs in the field of ICT support. When a software house supplies and installs a large, new bespoke system for an organisation the customer is likely to need considerable support for the first few weeks or months. New users will be unfamiliar with the package and may need guidance in its use. At this stage, unforeseen errors may occur in the software and modifications may need to be made. ICT professionals from the software house's development team can work on site with the customer organisation during the early stages of the software's use.

Qualities and characteristics required of an ICT professional ◀

The range of work in the ICT industry is huge and always changing. Each job will have specific skill and knowledge requirements. There are, however, personal qualities and general characteristics that are relevant to many fields of work within the ICT industry. It is often the possession of these, as much as specific ICT skills, that allows an individual to progress to a senior position.

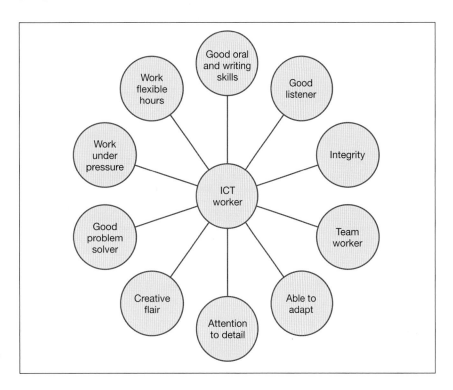

Figure 16.2 Personal characteristics needed by ICT professionals

▶ Good communications

The computer boffin, who rarely speaks to anyone but sits alone all day hunched over a computer screen, is a common public image of a person in the ICT industry. In fact there are few jobs that allow such isolation. Most require some communication with others – both oral and written. The jobs of some ICT professionals are described in case studies in the chapter. In all of them, the importance of good communication skills is emphasised.

Written communication

At all stages in the development of an ICT system, written reports are produced. These may include specifications, maintenance and end-user guides as well as progress reports for management. The ICT professional needs to be competent in writing a variety of documents in good clear language, in a suitable style, and using a level of technical detail that is appropriate for the audience.

Program developers need to produce written documentation for their programs so that, when a program needs to be modified at a later date, other developers can gain a clear understanding of what the program actually does.

A software developer, when investigating an end user's requirements for a new ICT system, needs to be able to make clear notes of interviews and provide the user with clear documentation of the system to enable discussion. In Case Study 4, Dave produces requirements specifications for his clients. To do this, he writes a report that brings together the findings of all his discussions and observations. It lays out, in detail, what the new system is to do. It is important that Dave writes the report in a clear, understandable way so that no confusion arises.

A member of a software company's help-desk team would need to be able to document faults clearly and document solutions using technical terminology for the use of a fellow professional. They would also need the ability to provide written instructions suitable for non-technical system users. Such instructions would need to be written in clear, plain English avoiding technical terms and jargon.

Oral communication

When working with end users of a system, ICT professionals need to be able to discuss problems with users and provide solutions in a clear, jargon-free and friendly way. This will enable efficient and effective communication. Considerable time may be spent in face-to-face discussion with end users who have little knowledge of the technicalities of ICT systems. If an ICT support worker does not have good oral communication skills, he is likely to confuse a user when he tries to help them solve their problems. Care must be taken in such circumstances not to use language that will antagonise, patronise or confuse; jargon should be avoided.

A member of a help-desk team will need to find out an end user's problem through careful questioning and then give instructions in a clear and understandable way. Case Study 2 describes how Jeremy has to feed back to a user the solution to a printer problem. He has to do this in a way that is clear and appropriate. In Case Study 5, Sam needs to discuss the requirements of a new system with the client in a way that does not overwhelm the client with technical ICT terms that he does not understand.

▶ Ability to listen

An ICT professional investigating an end user's requirements needs to be a good listener who concentrates fully so that he obtains a clear understanding of what the user requires. Many new systems have proved to be unsatisfactory because the ICT specialist produced what he thought the user should want rather than what was actually needed. Dave, in Case Study 4, is very careful to listen to the end user so that the system requirements that he produces fully meet the user's needs.

If a member of a help-desk team does not listen very carefully to the end user then the user's wants can be misinterpreted or ignored and inappropriate instructions can be given. This can lead to frustration for the end user; he may decide not to ask for help again.

▶ Integrity

An employee in the ICT industry must be trustworthy. An individual might have access to sensitive information and it is important that they can be relied upon not to misuse it.

A systems developer working with a client obtains detailed information about the organisation that needs to be kept confidential.

▶ Team working

The ability to work effectively in a team is essential for nearly all ICT practitioners as few work in isolation. A good team member is sensitive to the needs of other members, reliable, supportive and co-operative. In a strong team, ideas, views and information are freely shared and members build on the strengths of others.

▶ Getting on with a variety of people

ICT professionals are likely to need to interact with a range of people in an open and non-threatening way. When a member of an ICT support team goes to a user to help with a problem the user may be angry or frustrated. The support worker must display patience and act tactfully, politely and supportively. A software developer, when investigating the end user's requirements, needs to be approachable and gain the user's trust. If the user does not feel at ease it is unlikely that the developer will obtain all the information that he needs.

When a member of a help-desk team is attempting to solve a problem for a user it is important that she keeps calm and prevents the end user from getting flustered.

▶ Adaptability and the willingness to learn new skills

The one certainty in the ICT industry is that nothing stays the same! The rate of change in hardware and software performance has been very rapid in the last decade. Skills regularly need updating and old ways of doing things are often abandoned. A successful ICT practitioner must be able to adapt easily to new working methods. Much work is project based and an ICT worker is likely to move between teams. As projects overlap, it is not unusual to be a member of more than one team at the same time.

An openness to new ideas is essential. Sometimes training courses are available for updating skills but the most effective employee is one who is able to acquire new skills in a variety of ways. Jeremy (Case Study 2) has to keep up to date with the regular developments that occur in the college network systems.

A help-desk operator needs to keep up to date with changes in software.

▶ Thoroughness and attention to detail

There are many ICT jobs that require precision and a detailed approach. For example, if a programmer does not follow the specification exactly then the resulting program may have unpredictable effects when implemented. A system tester must ensure that every test is carried out as specified and that results are accurately recorded. Those involved with entering data into a live computer system need to work with a high level of accuracy as incorrect data entry leads to incorrect information.

▶ Creative flair

The ability to think innovatively and come up with new ideas is needed in some ICT jobs. An outstanding programmer needs to have more than the necessary technical language skills. To solve some problems, an effective programmer needs to think of new ways of finding solutions. However, programmers are likely to have to stick exactly to a given specification and are expected to work within tightly defined parameters. Creative flair cannot be used as an excuse to implement a system in a way that appeals to the programmer but does not meet the needs of the user as laid out in the requirements specifications!

If a web designer is to create exciting and effective sites, she will need to possess a strong visual sense and good spatial awareness as well as sound technical knowledge. Neera, in Case Study 3, needs to show such flair if she is to come up with solutions to unusual problems.

▶ Problem solving

An ability to approach problem solving in a systematic and logical way is essential for many ICT roles. Many organisations require new employees to take aptitude tests to demonstrate that they can take this approach.

A network manager may have to work out the cause and location of a fault in a network. To do so efficiently requires a systematic and logical approach. A member of a help-desk team needs to consider alternatives and find solutions to end-user problems. They need to stick at the problem and see it through until a solution is found.

When producing a requirements specification or designing a new system, an ICT professional needs to use an analytical approach to make sure that all factors are taken into account. Such an approach is necessary for Jeremy, Neera and Dave in each of their jobs (see Case Studies 2, 3 and 4).

▼

▶ Ability to work under pressure

There are likely to be many situations when an ICT professional is under pressure: a deadline could be looming, a program might not be performing as it should, a crucial hardware device could fail or a user could come up with unexpected and urgent demands. The pressures might mean that the employee has to work very long hours for a period of time.

The professional has to be able to take orders, be responsible for their own job and perhaps manage several different jobs at the same time. The ability to organise time and to prioritise tasks is vital. It is important that deadlines are met and work is not left to the last minute. This is apparent in Dave's role (Case Study 4). He has to meet the demands of several projects as well as putting in many extra hours of work to ensure that deadlines are met.

▶ Willingness to work flexible hours

In certain jobs, it is essential that the employee is able to work flexible hours. Many systems for multi-national organisations are worldwide and different offices are in very different time zones. An employee providing software support could be working in London, whilst the users are located in Los Angeles. When a new system is introduced, it is likely that support personnel will be required to work late into the night so that they are available to answer queries that occur during the Los Angeles working day.

An ICT support worker is often required to be available "on call". This means that for certain hours, when the employee is not in the office, she must be available to be contacted by pager to deal with queries. This could occur in the middle of the night. As an example, in Case Study 2, every fourth week Jeremy has to work until 8 p.m. on week nights and on Saturday to ensure that support is available to the users of the network.

When a network fault occurs a team member needs to find a solution as quickly as possible even if this means working past the official end of the day. New software may need to be installed at times when the usage of the network is low, perhaps in the evening or at weekends.

▶ Skills and knowledge

Each job has its own technical skills requirements. These are not personal skills. (*Remember to read exam questions carefully – do not include an answer about skills and knowledge if the question asks for personal characteristics.*)

Skills requirements might relate to the characteristics of specific hardware, the use of a range of facilities offered by particular software, or perhaps the knowledge of a given programming language. Such skill requirements are always changing and an ICT professional needs to regularly update their skills. A network manager needs to have very detailed and specific technical knowledge of networking. A web developer needs completely different knowledge and a different set of skills.

For ICT roles such as systems developers, a knowledge of general business practice is essential as well as a detailed understanding of the specific industry, such as banking, retail or education.

Activity 2

Imagine that you are preparing your CV to use in applications for jobs in the ICT industry.

1. Prepare a list of your ICT skills and knowledge.
2. Itemise the personal qualities that you possess which would show that you are suitable for employment. Back up each quality that you list with evidence to support your claim. For example, you could be good at team working as you have successfully taken part in an expedition for a Duke of Edinburgh Award.

Activity 3

Use the Internet to search for job advertisements in the following fields:

1. Database development   Project management
   Data entry clerk   Business analysis
   Network administration   User support
   Web development   Customer relationship management

2. Draw up a table with the following headings to display your findings:

Job title	Skills and experience	Personal qualities

case study 2
▶ Jeremy – ICT technician

Figure 16.3 Jeremy

Jeremy is the senior ICT technician in a further education college and has three technicians working with him. The normal working day is 8.30a.m. to 5p.m. but once every four weeks he has to stay until 8p.m. on Monday to Thursday and cover on the Saturday. The following week he has Monday off. Some tasks, such as adding a network printer to a classroom, which involves installing the printer driver in all 20 computers, can overrun at the end of the day. The task cannot be left as the computers will be needed by a class at 9a.m. the next day. Jeremy's job is varied. When he gets to college, his first task is to check that nothing has gone wrong overnight and that the backup procedures took place satisfactorily. Part of the day is spent staffing the help desk. Between calls he gets on with other tasks such as monitoring the network usage, studying the documentation for new software, or preparing material for a staff training session.

At the help desk, all user problems are logged on a database and when a problem is solved the resolution is added. Often a problem requires Jeremy to go to the user to resolve the fault on site. There are over 300 members of staff with vastly differing ICT skills. Jeremy has to make sure that the explanation and help he gives are appropriate to that user and that he neither goes over their head nor patronises them. For example, it is important that someone who reports a printer fault is not made to feel stupid when it turns out that they failed to check that the mains lead was attached. On the other hand, they need to be made aware so that they won't call the help desk for the same problem again!

The software and hardware used in the college is regularly updated. Jeremy has to teach himself the details of the changes. Sometimes the manufacturer may run a course, but more often he learns through help files available on the Internet and by reading articles in specialist magazines.

Much of Jeremy's time is spent on housekeeping and other network jobs. To enable him to do this he has full access rights to the entire network.

- Describe the personal characteristics that Jeremy needs to be successful in his job and explain why each is needed.

Sample answer: *Jeremy has to be prepared to work flexible hours as jobs such as installing a printer driver may require him to work past the end of the working day. His work on the help desk requires excellent oral communication skills as he works directly with end users. He needs to provide them with solutions to their problems in a suitable and understandable manner. He needs to be a good listener so that he can correctly interpret what a user requires and must be patient when listening to users' problems. Jeremy must adopt an analytical approach when solving a user's problem that has not occurred before so that a solution can be found as quickly as possible.*

case study 3
▶ Neera – technical consultant

Figure 16.4 Neera

Neera is a technical consultant for a business specialising in producing software for conducting e-business over the Internet and using wireless devices. She has to use her extensive technical knowledge to provide a detailed solution for a user's application and hardware. This requires a clear, logical approach to solving the problem, so that all factors are taken into account, combined with a creative flair that allows her to find ways to deal with unusual problems. She needs to present her solution in terms that the user will understand.

When Neera first meets a customer, they can be in a state of panic as she is often called in to resolve a live problem that is affecting current service. To reassure them that she can solve the problem, she needs to ask questions before the meeting and carry out some background research. She has to ask clear and concise questions and not get lost in too much detail. She needs to develop a rapport with the client and must establish some common ground and avoid disagreement.

Meeting with clients is only part of Neera's job. She also has to oversee the development of new software, resolve performance issues, discuss hardware requirements and provide plans for the next release of software. Many of the tasks she needs to complete stretch over a number of weeks. Each week she has to decide how much attention is needed by each task. Sometimes she has to work long hours as several tasks may require attention at the same time. In general, she works on her own with her client.

Neera finds consulting an exciting and diverse career but does not enjoy the travelling that it involves. Her daily travel varies from an hour on the train to two hours in a car, with the additional complication that her location from week to week, and sometimes day to day, is not known until the last minute.

■ Copy and complete the table below. The table should contain five personal qualities needed by Neera in her job, together with a justification of each.

Personal Quality	Justification
Needs to be able to get on with a variety of people	Neera's job involves her working closely with customers who have problems with their current ICT systems. She needs to gain their confidence to make sure that she finds out the information she needs as quickly as possible
Ability to work under pressure	
Adaptability	

case study 4
▶ Dave – ICT project manager

Figure 16.5 Dave

Dave works as a project manager in the central ICT department of a high-street bank.

After gathering user requirements, he designs the system solution and specifies the requirement for developers (programmers) on his team. He manages the implementation of the solution including the physical release of software. Knowledge of the business area and the system enables him to design the best solution for the users. He always has to consider the users' roles and how he can best solve their problems. It is very important that he listens carefully to what the user says by concentrating carefully and asking questions to make sure that he understands what he has been told. The new system needs to be based on what is really needed, not just what Dave himself thinks should be needed.

The initial stages of a project involve requirements gathering and the writing of specifications documented in a form that is clear to the users and the system developers. This liaison continues throughout a project, with informal meetings with the users during development and implementation.

In the banking world, work tends to be driven by deadlines. As a project nears the time when it needs to be delivered, it is often necessary to undertake evening and weekend work. The system that Dave works on has global coverage so different time zones often require late and early conference meetings. He currently travels to Paris and New York as his two main projects are based there. Often new tasks arise that have not been planned for. (This happened recently when a new project was prioritised by someone very senior in the bank.) This causes difficulties because he either has to manage to fit the additional tasks in or tell other users that their deadlines cannot be met.

His team is currently providing post-implementation support for a completed project and providing handover support to those responsible for day-to-day support. He is also managing two further projects in parallel which are at different stages in the project cycle, as well as assisting on two other projects for which his specific expertise is required.

■ Describe the personal characteristics that Dave needs to be successful in his job and explain why each is needed.

case study 5
▶ Sam – Web developer

Figure 16.6 Sam

Sam works for a national company that develops e-commerce systems for small businesses. He has to discuss the proposed project with the client so that he can find out exactly what the client requires the new system to provide. As the client is unlikely to have a deep knowledge of ICT it is important that Sam can discuss the project without using technical language that the client is unlikely to understand.

Sam has extensive technical knowledge that allows him to develop the systems for the client. He has to be able to consider alternative solutions to the problem so that he can find the one that is most appropriate in the circumstances.

As well as developing new systems, Sam also maintains and supports existing systems, some of which were developed by other people. Maintenance can involve modifying the system when errors are found or adding new features that the client has requested. To allow for successful maintenance of any system that he develops, Sam has to produce written documentation that explains exactly how the system works.

1. Find out more about e-commerce (see Chapter 17 or try http://www.howstuffworks.com).

2. Describe three personal characteristics that are essential for Sam in his job of Web developer.

▶ Worked exam question

When producing a requirements specification for an ICT solution to a task, an IT professional needs to use certain personal skills. State two such personal skills and explain why each of them will benefit the IT professional when preparing the requirements specification. (4)

▶ **EXAMINER'S GUIDANCE**

There are many characteristics that you could choose from; make sure that you choose two that you can back up with a good justification. A number of questions similar to this one could be asked. You must be very careful to look at the particular context (in this case, someone is preparing a requirements specification) and make sure that your answer fits it – don't simply rush in with a general answer. In Case Study 4, Dave has to prepare requirements specifications for his bank.

▶ **SAMPLE ANSWER**

An IT professional needs to have good written communication skills as he has to write specifications documented in a form that is clear to the users and the system developers (see case study 4).

He needs to have good listening skills. To produce an accurate specification, he must listen to what the user says by concentrating carefully and asking questions to make sure that he understands what he has been told. The new system needs to be based on what is really needed.

▶ **EXAMINER'S GUIDANCE**

Other possible answers are given below:

- *Be approachable and gain the users' confidence*
- *Have good analytical skills*
- *Be able to communicate well orally*
- *Be a good team worker.*

Produce your own justification for each based on Case Study 4. You will need to refer to the main text.

Remember that the examiner is looking for correct use of technical language. For example, you need to write "good oral communication skills" rather than "he can talk to people". The examiner will also expect you to write in appropriate style. You need to write full sentences that are grammatically correct. You should take care with your spelling too.

The use and organisation of ICT teams ◀

A team is a small group of people who work together on a common task or group of tasks. A help desk is likely to consist of a team; a business may have an ICT support team that looks after hardware, software and networking. When a new system is to be implemented, a project team is carefully selected with the aim of completing the project or subproject.

Although each team member may be allocated a specific role, in a successful team the members work together rather than as individuals. Tasks are usually allocated according to the strengths of the team members; getting people to do what they are good at is usually appropriate.

However, there are times when a team member needs to take on tasks usually undertaken by another person – when that person is sick or on holiday – so flexibility and understanding of other people's roles are necessary. Complementary qualities are required within an ICT team, although good oral communication skills are required by all.

Most teams need to have a leader who will liaise with management on behalf of the team and oversee the team's work.

A successful team is likely to have most of the following characteristics:

- good leadership
- an appropriate balance of skills and areas of expertise, with suitable allocation of tasks
- adequate planning and scheduling of tasks
- adherence to agreed standards
- good communication skills both with users and within the team.

A project team will work to a specific goal and will only exist as a team for the duration of the project. A project team will need to have a member with the skills to monitor and control progress against the project plan, as well as to control costs.

▶ Leadership

A good team needs clear and consistent leadership. The team leader should have sufficient seniority to fulfil the role and should have adequate understanding of the job of the team and the ability to adequately and systematically monitor and control performance and costs.

The leader must be able to hold the team together. A good leader will bring the best out of the individual members and encourage cooperation and exchange of ideas. Tasks should be allocated to the members appropriately, so that every member is given work of which they are capable and which, if possible, will help develop their individual skills. In this way, each task will be completed in the best way possible.

Leadership is important because appropriate management will encourage the members of the team to work together in an organised manner and motivate them to ensure that all deadlines are met and tasks are performed to the highest standards.

Much of a team leader's time is spent managing other people. He needs to arrange meetings, ensure that team members have the resources that they need, follow up any problems that arise and ensure that all the necessary things are in place to ensure that the job can be done.

▶ Balanced team

In some situations, there should be a balance of skills between team members who could have different backgrounds, for example, in systems, in business operations or in technical fields. This is particularly important for teams that are involved with planning.

In other situations, it is appropriate for teams to be made up of similar personnel, each of whom performs a small part of the whole. For example, a team whose task is to develop a system would consist of a number of programmers. The team often includes members with a range of skill levels, from trainees up to highly experienced specialists. This allows for the development of less experienced team members who can be set tasks within their capabilities whilst receiving help and guidance from their more experienced colleagues. The experienced team members are assigned the most complex tasks.

▶ Planning

In a successful team, tasks are planned and scheduled. For example, in a development team of programmers, there will be a range of tasks to be carried out including coding different sections, testing and documenting. The tasks will be allocated to team members and appropriate deadlines will be set.

In a school ICT support team, there may be a range of jobs that need to be carried out during a particular day: installing new hardware, monitoring network performance and troubleshooting when problems occur around the school. It is most important that work is planned and tasks are adequately scheduled to ensure that all tasks are performed. Without planning some unpopular tasks might not get done. The time period for planning might be long, medium or short term.

▶ Adherence to standards

Standards are agreed, formal ways of carrying out tasks. They may be nationally agreed, published standards which many system developers use or standards that are agreed within an organisation or even within a project. Standards involve the use of **formal methods** for the development of information systems, rather than team members drawing up their own standards.

One of the benefits of using standard methods is that they allow a team member to pick up another's work easily in case of an unplanned absence.

The use of established standards also ensures that a professional or methodical way of working is used. For example, it is important that appropriate documentation is produced and kept up to date. By following set procedures, the team will ensure that nothing is missed by mistake.

▶ Communication

Good oral communication is essential within a team as the members need to discuss ideas, work done and problems that have been encountered.

case study 6
▶ Sophie – project manager

Figure 16.7 Sophie

Sophie works for a large banking organisation. She manages a team of six full-time members and runs several projects at a time. A recent major project involved splitting up a current database system into two communicating halves, one dealing with **front-office** (customer-related) operations, the other with **back-office** (internal administrative) operations. This was done so that front-office services could be provided on a different hardware platform.

The project was split into three main phases:

■ initial analysis, when data flow and user interface needs were established

■ development, when the program code was written and tested and the interfaces with other software were tested

■ new system installation.

Sophie worked out that the project would require 20 man-months to implement. Four members of her team were allocated to the project. Other freelance contractors with specific skills were also involved. Developers (programmers) who had specific hardware experience and knowledge of both UNIX and NT operating systems were required within the team. Analysts on the team needed both a good technical knowledge and a sound understanding of the banking business. More junior team members were employed in testing.

At the start of the project, Sophie drew up a project plan that highlighted what tasks needed to be done to complete the project and who should do them. Sophie also maintained a spreadsheet of all tasks where details of problems that arose were stored.

As the project evolved, she met regularly with team members to discuss progress, giving help and direction as appropriate. If any problem became apparent she would provide extra resources to the task. She was able to monitor and control overall project progress as she could see when individual tasks were complete.

Whenever an unforeseen problem was thrown up by testing, the job of resolving the problem would be delegated by Sophie to a team member. She oversaw the progress of the project and carried out many of the tasks, such as managing the testing process.

Four months after starting the project it had been successfully implemented.

1. Give the reasons that ensured that the project was implemented successfully.

2. How did Sophie maintain control of the project?

3. Why were junior team members allocated to testing?

As well as requiring specific ICT skills, professionals in the industry are likely to need some or all of the following personal qualities:

- good written communication skills
- good oral communication skills
- ability to listen
- integrity
- team working and getting on with a variety of people
- thoroughness and attention to detail
- creative flair
- analytical approach to problem solving
- ability to manage pressure
- willingness to work flexible hours
- adaptability and the willingness to learn new skills.

It is important to be able relate the personal qualities to particular jobs as some attributes may not be appropriate for some jobs.

For an ICT team to be effective, it needs to have most of the following characteristics:

- good leadership
- an appropriate balance of skills and expertise
- adequate planning and scheduling of tasks
- adherence to agreed standards
- good communication skills both with end users and within the team.

Questions

◀

1. An important part of the development of an ICT solution is the production of documentation for its users. Describe two personal skills that are needed by an ICT professional when producing user documentation. (4)

2. A company wishes to recruit a new member for its ICT help-desk team. You have been asked to write a job description for the role. Explain two personal qualities that are relevant to the job which you would ask for in the description. (4)

3. Many retailers are expanding their e-commerce operations.
   a) Identify **one** current job for which ICT professionals are being recruited within e-commerce. (1)
   b) All ICT professionals require certain personal characteristics in order to work effectively. Using the job you have suggested in (a), give **two** personal characteristics for the ICT professional to work effectively and explain why you consider these characteristics would be essential. (4)

   AQA Sample ICT02

4. A systems analyst developing a new ICT system has to work with clients who have little, or no, understanding of ICT. State two personal qualities that the systems analyst should have that will enable him to work with the clients effectively and give an example of when each quality would be needed. (4)

5. A good leader is a major factor of a successful ICT team. State three further characteristics needed by a successful ICT team and for each explain why it is needed. (9)

17 ICT networks

AQA Unit 2 Section 4 (part 1)

◄

New developments in digital communication and networking are happening very fast. Many of the uses that we make of ICT today were not available just a year or two ago. It is important that you keep up to date with developments as they occur. One way that you can do this is by accessing the appropriate section of an online newspaper or industry-based online magazine such as:

- The Science and Technology section of the Independent – http://www.independent.co.uk/
- The Tech and Web News section of The Times – http://www.timesonline.co.uk
- The Technology section of the Guardian – http://www.guardian.co.uk/
- Computing – http://www.computing.co.uk/

Activity 1

Use one or more of the newspaper websites given above to find out about a very recent development in communications technology. Write a brief description of the development in your own words to share with the members of your class.

Working through case studies is very important for this topic; many are included in this chapter and the next. It is possible that you will be given a case study to read and answer questions about in your ICT02 exam.

Global communications ◄

All organisations and individuals need to communicate, to send information to and receive information from other organisations or individuals. ICT has transformed the way we communicate. New methods of communication have been developed, each with its own features and advantages, for example:

- email
- fax
- Internet
- Internet conferencing

- instant messaging
- mobile phone
- satellite phone
- Short Message Service (SMS), commonly referred to as text messaging
- videoconferencing
- viewdata
- Voice over Internet Protocol (VoIP)
- Personal Digital Organiser (PDA)
- intranet
- extranet.

Activity 2

Copy and complete the following grid.

	Description	Example of use
Instant messaging		
Viewdata		
SMS messaging		

Public networks such as the Internet, to which anyone can connect almost anywhere in the world, mean that millions of computers can be linked together. This means that users can:

- communicate with each other quickly, for example by email
- share files
- use browser software to access web pages
- search for information.

Activity 3 – using a search engine

Search engines, such as Google (the world's most popular search engine) and Lycos, enable users to search the Internet using selected keywords. A search engine is a program that allows a user to enter a query and will search a very large database to find matching items.

If you want to attract visitors to your site, it is a good idea to put keywords into the HTML script for a web page (see Figure 17.1) so that they are picked up by the search engines.

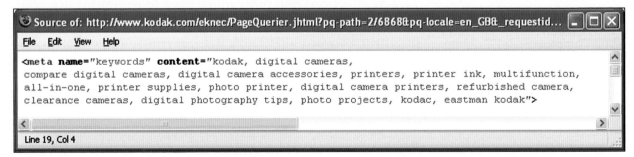

Figure 17.1 The keywords from the Kodak site

Activity 3 (contd)

1. Go to a search engine such as Google. Type the search condition: diy online. What do you get? Do you get a link to http://www.diy.com?

2. Visit a commercial site such as http://www.cadbury.co.uk. Select View > Source from the browser menu. Scroll down to the line beginning <meta name="keywords". Find the keywords for the site. Then go to a search engine and search on three of the keywords. Is the site listed?

Search engines such as Google allow companies to set up sponsored links related to specified keywords. The links are displayed prominently on the page (see Figure 17.2). The company is charged at an agreed rate for every time a user clicks on their link.

Figure 17.2 Google displaying sponsored links

3. Where are the sponsored links to be found?

4. Investigate the use and costs of sponsored links.

case study I
▶ using VoIP

Figure 17.3 Skype in use

Skype provides a free Voice over Internet Protocol (VoIP) service that allows a user to talk to other users anywhere in the world. To use the service, you simply need to download the software from the Skype website on to your computer and make sure you have a microphone and speakers or a headset plugged into your computer. If you want the person you are calling to see you during the call, you will need a webcam as well.

Anne has two sons: Tom lives in Australia and Dave in a different town in Britain from Anne. Anne, Tom and Dave each have Skype installed on their home computer. Anne is able to talk to Tom very regularly and appreciates being able to see him as they chat.

Dave has three young children, Anne's grandchildren. They all enjoy chatting on the phone and showing their grandmother their latest toys and paintings.

1. Describe the benefits of using VoIP.

2. What does a user need to be able to communicate with someone using VoIP?

3. Access the Skype website at http://www.skype.com. List other features that are offered by Skype.

Figure 17.4 Anne's grandchildren talk to her over VoIP

case study 2
▶ David using a PDA

Figure 17.5 David

David works for an international bank as a project manager. He has been issued with a Personal Digital Assistant (PDA) to help him keep in contact when he is away from the office. A PDA is a hand-held device that functions as a mobile phone, a personal organiser and a web browser. It is also able to send and receive email messages.

David works closely with offices in New York, Tokyo and Paris. He has to be available to make quick decisions and offer advice whenever problems occur. The best method of communicating with his colleagues abroad is by email. Using his PDA, he is able to access his messages when he out of the office, whether at home or when he is travelling.

1. Why is email a better method than telephone for communicating with colleagues in New York and Tokyo?

2. What drawbacks are there for David in having a PDA issued by his employer?

3. Research and describe other functions offered by PDAs currently on the market.

4. Can you think of any further functions that could be added in the future to improve the use of PDAs?

Activity 4 – mobile phone features

1. Make a list of features that are available on mobile phones, such the ability to take photographs. Think of as many as you can.
2. Ask 10 members of the class (or your friends) whether they have a mobile phone with each feature. If it does have a feature, do they make regular use of it? If it does not have a feature, would they like to have it?
3. Count up the responses that you have and create a table such as the one below:

Feature	Number having feature	Number regularly using feature	Number who would like feature
Ability to take photos			

case study 3
▶ the growth of IPTV

TV over broadband Internet (IPTV) is a growth area in UK broadcasting. TV programmes are made available over broadband. A viewer can see a programme that they have missed whenever they choose. TV shows and films can be downloaded to a PC or to a TV via a set-top box. The viewer can rewind, fast forward, pause and control the programmes in the same way as they can a DVD.

BT Vision is a digital TV service that is available to homes in the UK. BT Vision offers digital TV (Freeview via a TV aerial) and a digital video recorder (called a V-Box), and allows TV and video to be downloaded over a BT Broadband connection.

1. What is Freeview?

2. What benefits does IPTV have for a viewer?

3. Give some examples of situations when an individual could benefit from IPTV.

4. Explain in your own words what a home user needs to access IPTV.

Many schools and colleges have installed **Virtual Learning Environment (VLE)** software. A VLE allows students and their teachers to interact online.

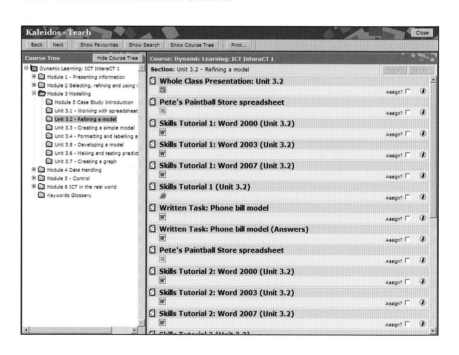

Figure 17.6 A VLE in use

A VLE offers the following features:

■ Access to course material and learning resources. Materials should be varied and include course content as well as quizzes and a range of exercises. Teachers can develop these materials using straightforward templates. A student can work through the material at his own pace, receiving feedback to reinforce understanding.

- Methods of tracking a student's progress in working through online material and assessment marks. Each student is able to check his or her individual progress; the teacher has access to the progress of the whole class.
- Means of communication between the student and his teacher that can provide feedback on his progress. This could be in the form of email or online chat.
- Means of communication with other students on the course using discussion forums and chat rooms.

In most cases, a VLE is an additional resource for students and teachers; it does not replace face-to-face contact in the classroom.

Network environments ◄

A computer network consists of two or more computers and peripherals that are linked together. These links can be made with cables. A **wireless network** uses radio waves, in the same way as televisions, radios and mobile phones, to make the links. A wireless network following a common protocol is called **Wi-Fi**.

Wireless networks are used in many homes. Such a wireless network is easy and inexpensive to set up. Home Wi-Fi can be set up so that several members of the household can access the Internet at the same time. As the name implies, no wires are required. In fact, it is possible to access the Internet using a home wireless network while sitting in the garden! It is very important that access is restricted to authorised users when the network is set up.

Wireless networks are also made available in places such as hotels, libraries and airports for general use. They are known as **Wi-Fi hotspots**. The greater the number of Wi-Fi hotspots that are available, the easier it is for someone with a laptop to keep up to date and in touch while on the move.

When we consider the elements and uses of a network we need to differentiate between the network needs of an individual user and that of an organisation with a number of users. With a large network, issues such as access to the network, security, data sharing and maintenance need considerable management; typically, a network manager with a team of technicians is employed to carry out this work. Such a network can be called a managed network.

There are two main types of network organisation:

- server-based
- peer-to-peer

▶ Server-based networks

Server-based networks have a central computer called a **server**. Other computers in the network are called **clients**. Devices are treated as either servers or clients; they cannot be both. Server-based networks are usually used by large organisations.

Servers provide central services such as backup and software installation as well as providing centralised storage of data files. This central storage provides a pool of data that is accessible to all workstations on the network.

Clients send requests for services, such as the retrieval of data or printing, to the appropriate server which carries out the required processing. Some processing tasks are carried out by the server and others on the client computer.

As software is installed centrally, it only has to be installed once. Backup is easy to perform and there is no need to rely on users backing up their own files.

Individual users can be set up centrally with appropriate access rights. Each user is allocated a user name, a password and disk space on the server. Security is therefore very high.

Large networks may have more than one server. The performance of this sort of network is heavily dependent on the servers. They need to have fast processing speeds, and a large amount of main and backing storage. They are therefore relatively expensive and server-based networks can be complicated to install. If a server fails then the clients on the network cannot access the resources provided by the server.

▶ Peer-to-peer LANs

A peer-to-peer Local Area Network (LAN) is a very simple network that provides shared resources to the computers that make it up. All the computers in the network have the same status and there is no central server.

Computers on a peer-to-peer network can access files and devices on another linked computer providing the appropriate access privileges have been set. Any computer can communicate and exchange data with any other on the network. Such a network cannot have complete security.

Installing software takes more time, as it has to be installed on each station. Backing up must also be done separately for each individual station.

As a server is expensive to buy, a small peer-to-peer network would probably be much cheaper than a small server-based network. A peer-to-peer network would be ideal in a small office where four PCs need to be networked to share data.

Both types of network can share printers and other peripherals such as scanners. Both types of network can be used to send and receive emails from other network users.

Many home users use the Internet in a peer-to-peer mode – when using personal publishing tools to share their videos, photos and audio with friends across multiple websites, blogs and social networks for example. The photographs are not stored on a central server but are accessed directly from the originator's own hard drive. This removes the need for file uploads, but both the originator of the photographs and the viewer need to be online at the same time for the files to be accessible. Users can share a file almost instantly, without having to create or maintain an online gallery. For examples of personal publishing software, see websites, such as http://www.mixpo.com/ and http://www.photoshow.com/.

Elements of a network environment ◄

► Communication devices

To connect a computer to a local area network, a **network interface card** is needed. The card is an electronic printed circuit that fits inside the computer into an empty slot on the computer's **motherboard** (main printed circuit). It allows the computer to be recognised by the network and allocates it a unique identifying number. The network interface card determines the maximum speed of data transmission that will be available around the network.

Individual users

A number of devices are needed to enable an individual home user to link to the Internet.

A **modem** (modulator–demodulator) is a device that can be plugged into a computer or, more usually, is housed inside a computer. It is connected to a telephone line and provides access to a wide area network. The modem converts the digital signal that is used within the computer into the analogue form that is required for transmission over a telephone wire and converts the analogue signal to a digital one for incoming messages.

A **broadband modem** also connects a computer to the telephone line. It uses **Asymmetric Digital Subscriber Line (ADSL)** modulation technology and special compression techniques to achieve much faster transmission rates. Typically, a broadband modem links a single computer to the Internet and the user is able to stay online all day.

Wireless broadband modem **routers** can be used by subscribers to broadband. The router allows several devices to share the Internet connection wirelessly, as long as each device has a wireless adapter. The wireless adapter translates the digital data from the device into a radio signal. The signal is transmitted using an **antenna**. When a device sends data to the Internet, the router receives a radio signal from the device and changes the data from radio-wave form to digital form. The router is connected to a telephone socket with a cable and the digital data is sent via this to the ADSL line.

When data is received by the router from the Internet, the router translates it into a radio signal which it sends to the computer.

When a household installs broadband, the broadband signal is available at any telephone extension socket in the house as well as the main socket. Every phone extension that is in use needs a **microfilter** to cut out the broadband interference.

A connector card can be added to a mobile phone to allow to it connect to a Wi-Fi network.

A USB wireless adapter allows a user to enable a laptop without a built-in wireless adapter to connect to wireless networks. The adapter is compact and easy to carry, and can be swapped between a laptop and a desktop computer quickly and easily.

Large organisations

Extra hardware is required when linking computers together to form a network.

A large organisation will maintain a number of servers. These are used to manage the traffic around the computers in the network. Attached are hard disks with very high storage capacity that store databases, user files and software that is shared between users and downloaded to users' client computers.

A switch handles the efficient transfer of data between different sections of the network. A repeater may be needed to amplify the signal on long stretches of cable.

▶ Data transfer media

Cables are the most common form of transmission media used to connect network stations to the rest of the network. The network cabling connects to the network interface card. Cables are usually made of copper wire. Common examples are **Ethernet** cabling (**coaxial**, like a television aerial) and UTP (**Unshielded Twisted Pair**) cabling. **Fibre-optic** cables are becoming more common; they are faster than copper cabling but more expensive. Different types of cable have different physical capacities; this affects the speed at which data can be transmitted around the network.

▶ Networking software

Special network operating system features are required to enable the running and management of a network. It sets up the correct protocols to be used across the network. A **protocol** is a standard set of rules that defines how communications take place between computers. Most modern operating systems, such as Microsoft's Windows Vista, include networking features as standard.

The use of protocols means that a network does not have to be restricted to one manufacturer's equipment; it allows for the existence of what are known as open systems. This means that several disparate pieces of equipment can be connected together and can communicate effectively. A network could have some laptops, some PCs and some PDAs that all intercommunicate.

A **modem driver**, a program that controls the broadband modem, will be required.

A **browser** program is needed to view web pages written in HTML. There are several examples, such as Microsoft Internet Explorer, Netscape Navigator and Mozilla Firefox. A browser allows users to retrieve information from the Web interactively over the Internet. It provides facilities for a user to store the addresses of commonly visited sites as bookmarks or favourites. It stores pages locally on the computer so that pages load quickly if they are revisited.

The user needs to subscribe to an **Internet Service Provider (ISP)** such as AOL, Tiscali or Wanadoo. The ISP normally provides an email address to enable the user to send and receive email and a limited amount of web space so that the user can set up a website. The ISP has a host computer that deals with communications and stores data such as email messages and web pages for the user.

An ISP provides users with a software package that they need to install on their computer. This enables the user to log on to the Internet and make use of the available facilities such as the Web and email. The software may offer the facility to set up filters so that certain types of website are blocked for users. Many parents use this as they wish to prevent their children's access to undesirable sites.

In a large managed network, special network software, such as Novell Netware, is needed to manage the many user accounts. For example, it is important that a network offers different users different levels of access. The network manager has unlimited access to all areas and drives. The manager needs greater privileges than ordinary users, in order to install new software, add and delete users, set up menus and so on. The manager's programs must be protected by passwords.

► Standards and procedures

A **standard** is a common way of doing something, for example, storing data in a particular format or transferring data in a predetermined way (see Chapter 18).

Without standards, networking would be limited to communicating between computers of the same type. The use of standards allows each device to interpret correctly the data that it receives.

Many **procedures** need to be carried out if users are to access any network safely and easily. These procedures only have to be undertaken because of the use of the network – they do not in themselves carry out any user task.

It is vital that every user of a public network installs a virus scanner. A **virus scanner** is software that is used to check data entering the computer to ensure that it does not contain any virus. The user must install updates for this software very regularly as new viruses are being written all the time.

A user should also install and maintain software to protect his computer against spyware. **Spyware** is software that secretly gathers information about a user through his Internet connection. Such software can secretly be installed on a user's computer when he downloads and installs a program from the Internet. Once installed, the spyware monitors user activity on the Internet and transmits that information to someone else. Some spyware is able to collect email addresses, passwords and credit card numbers.

Many software manufacturers issue maintenance modifications to their software packages as downloads from the Internet. A user should ensure that these are carried out. In a large, business network, a number of regular procedures need to be carried out. These include:

- Maintaining software to the network: new software must be installed on the appropriate server and maintenance modifications made from time to time
- Adding, deleting and modifying user accounts; this involves aspects such as access rights, memory allocation and monitoring use
- Carrying out regular backups
- Monitoring of network traffic and usage
- Maintaining security across the network.

The Internet and the Web ◄

The **Internet** is a very large number of computer networks that are linked together across the world via telecommunications systems. Messages and data are sent from the source computer, through a number of other computers until the destination computer is reached.

The **World Wide Web (Web)** is a vast collection of pages of information in multimedia form held on the Internet. Pages can contain images, videos, animations and sounds. An organisation or individual can set up a website consisting of stored pages that are made available to other users. Much of the material on the Web is freely available to anyone. Websites have a home page which provides links to other pages within the site.

Some pages are password protected and are only available to subscribers. Businesses may password protect some of their information pages so that they are available to their employees but not to the general public.

Web pages are written in a language called **HTML (Hypertext Mark-Up Language)**. Web pages can be created and websites built using web design software such as Microsoft FrontPage, Expression Web or Adobe DreamWeaver or just written in HTML using a text editor such as Notepad. The user can create, edit or delete pages and set up or edit links between pages to allow easy navigation of the site.

▶ Accessing the Internet from home

Once connected to the Internet, the user can:

- access pages of information on the Web, including text, images, sound and video
- save Web pages and images locally for later reference
- leave messages on 'bulletin boards' and join discussion forums
- share user videos (on sites such as YouTube)
- talk to friends on a social networking website, such as Facebook
- send and receive email.

The ISP may also provide:

- free web space to set up and edit your own web pages
- additional email addresses for the user's family
- latest news, weather, TV and radio information
- its own search engine for searching the web
- its own Internet shopping facility
- bulletin boards for newsgroups (special storage space which is used for messages relating to a particular interest group, for example, Star Trek, old computers, coarse fishing or teaching ICT).

Activity 5

1. Write down the facilities from the lists above that you have used.
2. Compare your answers with other members of your class.
3. List any further features that you have used on the Internet.

▶ Other ways of accessing the Internet

Big organisations where a large number of users need to access the Internet are likely to find that a modem and dial-up connection, and even ADSL broadband facilities, will not meet their needs. The users' computers are likely to be organised into a local area network which is linked to the Internet via a terminal server. The organisation is likely to have a permanent link to the Internet via a fibre-optic cable.

Televisions with digital capabilities, either digital TV or a normal TV with a conversion box, allow a user to access the Internet.

Mobile phones and personal digital assistants (PDAs) that can access the Internet are becoming more common. Obviously screens are small but otherwise web pages appear little different from on a PC. Navigation is carried out by pressing arrow keys and using another key to make a selection.

Web addresses can be typed in using a small built-in keyboard or an on-screen keyboard. Standard email facilities such as reading, replying and forwarding messages are also provided. You can attach files as well.

Wireless "hot-spots" where laptop users can access the Internet are increasing in number. The laptop needs a wireless network card. In many places, such as coffee shops, this service is free.

Activity 6 – Comparison websites

A number of websites exist to provide users with the opportunity to compare prices of goods or services of products from different suppliers. Uswitch.com is one example, which states the following:

USwitch.com is a free, impartial, online and phone-based comparison and switching service that helps customers compare prices on a range of services including gas, electricity, home phone, broadband providers and personal finance products. Our aim is to help customers take advantage of the best prices and services on offer from suppliers.

1. Explore Uswitch on **http://www.www-uswitch.com**. Find out how much your family pay for one or more of the services. Using the website, find out if your family could save money by switching to a different provider.
2. Another price comparison site is **http://www.kelkoo.co.uk**, which allows a user to find out the price of a product, such as a DVD, from a large number of online suppliers. Explore the range of prices for an item of your choice.

▶ The Internet and business

With hundreds of millions of people across the globe using the Internet, commercial interests have recognised the opportunities offered by this new technology.

The Internet offers businesses many opportunities. For example, they can:

- market their products to a worldwide audience with a website
- carry out research
- sell products directly over the Internet (e-commerce)
- use videoconferencing for virtual meetings
- use Electronic Data Interchange (EDI) to communicate with suppliers
- use intranets and extranets to get up-to-the-minute information
- use email to communicate quickly.

Electronic commerce (e-commerce) is the term used to describe the conducting of business transactions over networks and through computers. It is usually used to describe buying and selling goods over the Internet but includes other electronic business transactions, such as Electronic Data Interchange (EDI) – where information such as orders and invoices are exchanged electronically between two companies – electronic money exchange and point-of-sale (POS) terminals.

▶ Commercial websites

Many businesses use websites to promote themselves. Some have found that not only is it much cheaper than conventional advertising but it reaches a much larger audience. They can also include animations and videos to make the site more attractive. It has been predicted that online shopping will account for a quarter of all sales by 2010.

When a company sells goods or services online it can offer the customer a great range of items. If a company such as Amazon were to sell their goods traditionally in shops they would only be able to offer limited stock in any store. If they were to set up a mail-order business, they would be unable to produce a catalogue of all the items they sell – a catalogue would have to be massive and would be much too expensive to produce and distribute.

Selling online is particularly appropriate for sellers of specialist goods. It allows them to locate their premises anywhere they wish and still have a worldwide potential market.

To be able to sell goods online to new customers, a company must first get the customer to visit the site. The Internet offers a number of facilities that the company can use to encourage use of their site by directing customers to it. The company can:

- register with a search engine
- place an advert, a popup window or a link on a related site
- add a meta tag to the web page – to provide information such as who created the page, what the page is about and the keywords for the page – that can be used by search engines
- add a function to their own site so that visitors can add the email address of friends.

Online shopping is particularly useful for customers who are housebound or for those who live very busy lives. Tesco, like several other food retailers, runs a successful online ordering service. Customers can choose their goods from their computer at home. They can select a convenient delivery time and are therefore saved the time and bother of travelling to a store.

Not everyone is able or willing to purchase goods online. To be able to do so they would need to have a computer with Internet access, ideally using broadband. Many people are hesitant to make payments online as they are concerned about security issues and are unwilling to give their credit or debit card details as they fear that they might become the victim of fraud.

case study 4
▶ AquaPress

AquaPress is a company that sells specialist diving books. It was established in 1996, when it sold its books by telephone and post. They produced catalogues of their products; these were costly and time consuming to prepare and distribute and constantly needed updating.

They established a website which now provides information on all their products and allows customers to order online. As they use broadband, their site is available at all times. The number of orders is now at such a high level that the use of broadband technology is crucial.

Since they started selling online using broadband technology, they have been able to improve their service to their customers in a number of ways. They now send out a weekly email which describes their latest products to all their subscribers, currently around 10,000 people.

The website also allows a customer to download sample chapters so that they can decide whether or not they wish to buy the book. This has resulted in a large increase in the sale of certain books.

1. Investigate AquaPress's website at http://www.aquapress.co.uk. What features can you find?

2. List the benefits to AquaPress of selling online. Include those benefits discussed in the case study as well as any others you can think of.

Intranets and extranets ◀

▶ Intranet

An intranet is a network based on Internet protocols that belongs to an organisation. It is made up of web pages which can be accessed by standard Internet browser software such as Microsoft Internet Explorer or Mozilla Firefox. The intranet is only accessible to employees of the organisation and must be accessed using an identity code backed up with a password.

An intranet website looks and behaves in the same way as any other website. It can be used across a local area network, a wide area network or normal Internet lines, when it is protected from unauthorised access by a **firewall**.

An organisation uses its intranet to provide employees with information. For example, letters, documents, schedules for the day, stock information, orders due for delivery and weekly sales figures can be made available on the intranet. The web pages can be accessed by authorised users over the Internet so they allow employees based at different locations worldwide to share a wide range of private data without the need for establishing a private network. Developing an intranet is a much cheaper option. An intranet can provide a shared diary system as well as an internal email system.

When the use of an intranet is well established in an organisation, the volume of paper documents that need to be distributed can be greatly reduced. An in-house phone book or health and safety manual will no longer need to be printed and distributed to every employee but can be accessible on the intranet. Not only is money saved but these documents can be kept up to date; changes can easily be made when they occur – it is not necessary to print out a new version of the document.

As the intranet web pages can be accessed using a standard Internet browser, the intranet can be accessed by users with any type of computer hardware. An intranet is easy to install and easy to use as the user is able to access its pages using familiar browser software.

Software is readily available to build an intranet without requiring a large team of programmers to create bespoke software for the organisation.

There are a number of issues relating to the establishment of an intranet within an organisation. How well these issues are met will determine its success and volume of use.

Care must be taken when devising a new intranet to develop an appropriate house style that will provide a consistent look and feel to all pages. Considerable thought must be given to the structure of the intranet and how specific information can be accessed. The task of keeping the information up to date must be assigned to the appropriate person or people. Out-of-date or incomplete information will quickly result in dissatisfaction in the user.

Adequate training must be given both to users and to those who have been allocated the task of updating information.

▶ Extranet

An extranet is an intranet that is made partially accessible to people outside an organisation. These people must be authorised. An extranet can only be accessed by someone who has a valid username and password. The user identity will determine exactly which parts of the intranet can be viewed by the user.

Extranets are widely used and provide an effective and secure way for businesses to exchange information. Typically, a business might share part of its information with suppliers, customers, or other businesses. A supplier could access sales information for their products. Having this up-to-date information will help them plan their production schedules. Customers could have access to details of their past and current orders; they might be able to track the progress of any outstanding orders.

Two companies could use an extranet to exchange large volumes of data using **Electronic Data Interchange (EDI)**. For example, stock orders could be transferred in this way to suppliers.

Online catalogues of products can be shared with selected customers. For example, a clothing manufacturer may make their catalogue of current items available to the wholesalers with whom they trade.

A school or college could set up an extranet by making part of their intranet available to parents of pupils so that they can access selected pages to obtain information relevant to their son or daughter.

It is important that information is kept secure when being accessed. When extranet links use the Internet, data can be encrypted or privately leased secure lines can be used.

case study 5
▶ the Bullring intranet

Figure 17.7 The Bullring shopping centre in Birmingham

The Bullring shopping centre in Birmingham is one of the most popular shopping centres in Europe. The retailers share an intranet and each store can access information about the building, its services and facilities. The intranet gives the retailers easy access to site information when they need it and helps to create community feeling among the shopkeepers.

1. Explain the difference between an intranet and the Internet.

2. Explain how an intranet could be used by a school or college and describe the information that could be stored on it.

case study 6
▶ WHSmith extranet saves £1 million

WHSmith News uses an extranet to allow its key customers – such as Tesco, Asda and its own retail arm, WHSmith – to access sales data to improve their efficiency and help spot trends.

Richard Webb, business systems manager at WHSmith News, said the reports enable the company to react to sales trends and to help customers with their internal reporting to support decisions about production quantities. "It also shows how we're performing. It's much more of a collaborative approach."

WHSmith News saved £1 million in costs in its first year by using ICT to streamline its supply chain processes and reduce waste of the magazines it sends out to retailers.

1. What advantages, other than cost savings, could the use of the extranet bring?

2. What other ways could the extranet be used?

3. What information could be made accessible to employees of WHSmith on a company-wide intranet?

SUMMARY

New communications technologies are developing all the time. Individuals and organisations can communicate using **ICT** in a wide variety of ways.

A **network** consists of two or more computers and peripherals that are linked together.

A **wireless network** uses radio waves to link devices. A wireless network following a common protocol is called **Wi-Fi**.

A **server-based network** has a central computer called a **server**. Other computers in the network are called **clients**. Devices are treated as either servers or clients; they cannot be both. Servers provide central services for the clients which are operated by the users.

A **peer-to-peer network** is a very simple network that provides shared resources to the computers that make it up. All the computers in the network have the same status and there is no central server.

Basic elements of a network are:

▶ **Communication devices**, including a network interface card and a modem, a broadband modem or wireless broadband modem routers

▶ **Data transfer media**: cables are the most common form of transmission media used to connect network stations to the rest of the network

▶ **Networking software**: special network operating system features enable the running and management of a network; a modem driver; and a browser program to view web pages.

The **Internet** is a very large number of computer networks that are linked together around the world via telecommunications systems.

The **Web** is a vast collection of pages of information in multimedia form held on the Internet.

E-commerce describes the conducting of business transactions over networks and through computers.

An **intranet** is a network based on Internet protocols that belongs to an organisation.

An **extranet** is an intranet that is made partially accessible to people outside an organisation.

Questions

◀

1. Explain the difference between the Internet and the Web. (4)

2. Describe the basic elements of an ICT network. (8)

3. Home users access the Internet to search the Web for information and send emails to their friends. Describe other ways in which a home user might make use of the Internet. (10)

4. A British company has offices in California and New Zealand. Describe three ways in which the company can use the Internet to communicate with its offices abroad. (6)

5. A company that makes squash racquets intends to set up a website to advertise its range of products and to take orders online.
   a) Describe two ways in which the company could use the facilities available on the Internet to encourage visitors to use their site. (4)
   b) Describe the benefits to the company of taking orders online. (4)

6. A children's toy retailer has decided to set up an online store.
   a) Explain two advantages to the retailer of using this method of selling as opposed to selling from a high street shop. (4)
   b) Explain the benefits and disadvantages to the customer of using the online store rather than a high street shop. (6)
   c) Describe two ways in which the retailer could make use of the Internet to publicise its new service. (4)

7. Explain the difference between a client–server and a peer-to-peer network. (6)

8. A retail company has decided to set up an intranet for the use of its employees and enable certain information to be available on an extranet.
   a) Explain what is meant by the term extranet. (2)
   b) Discuss the advantages to the company of installing an intranet and the issues that must be considered. (10)

9. Changes in technology now mean that it is no longer necessary to have a PC to be able to use some Internet services. Describe other ways that Internet services can be accessed. (6)

18 *Use of communications technologies*

AQA Unit 2 Section 4 (part 2)

Use in the home ◀

A high percentage of homes now have one or more computers which are used in a variety of ways not related to a business. Examples include:

- keeping in contact with friends and family using email and VoIP
- accessing the Web for information and services
- online shopping and banking
- downloading, viewing modifying and sharing photographic images
- accessing social-networking sites (SNS), such as Facebook
- maintaining a Web log (blog), a personal journal that is made accessible over the Web
- instant messaging.

Figure 18.1 Facebook

▶ Social networking

A social-networking site allows a user to create online links with other users who are friends or colleagues. These links build up to create communities as a small initial group of users invite their friends to become part of their own personal network. These friends then invite others and so on. There are now hundreds of social-networking sites worldwide; millions of users have accounts.

Users share basic personal information including a photograph, as well as information such as their favourite music, books and films to establish their identity. They keep in touch with friends, exchange information and share links.

Additional software can be added that offers a peer-to-peer service that allows a user to share or stream media such as music to their friends.

There has been a recent growth in the number of small, home peer-to-peer networks (see Chapter 17). These are very easy to set up and use.

Activity 1

1. Choose three friends or neighbours who have a computer with Internet access. Try to select people from different age ranges. Ask each person to keep a log of what use they make of their computer during the course of the week. Compare the three logs.

2. Prepare a tutorial using presentation software on the use of social-networking software.

3. Discuss the benefits and problems of the growth of social-networking software.

▶ Connecting to the Internet

A home user who makes little use of the Internet can access it by using a **dial-up** facility. In this case, the home computer is connected to the Internet through a modem and can only operate at the speed of the modem, typically 56 Kilobits per second (Kbps), although ISDN digital telephone lines offer faster speeds of 128 Kbps. While the modem is using the telephone line, a user cannot make a standard phone call while they are online. Customers are charged for the amount of time that they are online. Such access is suitable for very occasional Internet users as the cost of access is relatively low; in some locations, dial-up is the only access method available.

In practice, many home users find that dial-up access is too slow to meet their needs. Demand has grown for greater bandwidth and faster Internet access. Many companies now offer broadband technology using fibre-optic cables giving speeds of at least 4 Megabits per second (Mbps). Using a broadband connection allows a user to be always online; connection and downloading time is included in the rental cost.

Asymmetrical Digital Subscriber Line (ADSL) technology has been developed to give broadband performance using a standard copper telephone cable. Special hardware and efficient compression techniques mean that ADSL can operate at high speeds. It is called asymmetrical because it downloads (receives data) more quickly than it uploads (sends data). The availability of ADSL to a user depends on where he lives. The household needs to be connected to an exchange that is enabled for ADSL and located within a few kilometres of the exchange.

As most Internet users receive a lot of data and only send emails, ADSL is an attractive option and cheaper than dedicated fibre-optic cables. It is not suitable when a user needs to send data at high speeds, for example, if hosting a website.

ADSL broadband access to the Internet incurs a payment of a monthly subscription. Different speeds and maximum monthly downloads are available at different costs. ADSL broadband access allows the telephone line to be used for voice calls at the same time as accessing the Internet.

Cable access is currently available in many towns and cities. It is only available if the area has fibre-optic cable for TV installed; installation is expensive and is only viable in highly populated areas where many people will access the services. Similarly to ADSL, a cable modem is connected 24 hours a day, 7 days a week and the upload speed is slower than the download speed.

Wireless broadband uses secure radio signals instead of a phone line or cable connection. It is available in certain areas of the country, sometimes where other methods are not available. In particular, certain sparsely populated areas can benefit from access to wireless broadband as houses are too far from local exchanges to be able to access broadband through a telephone line. Some local groups in rural areas are building wireless networks to bring high-speed Internet access to themselves and members of their community.

case study 1
▶ BT broadband

At the time of writing, BT was making the following offer to encourage people to switch to broadband:

- Free modem (or go wireless with a free BT Home Hub)

- Free connection if ordered online

- Free PC antivirus and firewall software

- Parental control, pop-up blockers and 15 MB of website space

- Support for services, such as digital TV with BT Vision, cheap phone calls with BT Fusion and video calling

- Inclusive Wi-Fi minutes from BT Openzone Wi-Fi access points

- Technical support at local rates.

1. What is a "Home Hub"?

2. What is the role of firewall software?

3. What is a "pop-up blocker" and why is it needed?

4. What is Wi-Fi?

Business use ◀

A small peer-to-peer network that might be used at home or by a very small business is relatively straightforward to set up. However, large, commercial networks need to be set up very carefully. User accounts need to be set up and maintained with appropriate access rights. File servers for software and data and access to the Internet with adequate security measures need to be put in place.

Many establishments preload every computer in the network with the "image" required so that it operates according to the required policies. This means that users cannot add their own software to their local hard drive but must only use the software provided by the organisation.

case study 2
▶ a laptop on a network

Figure 18.2 Natalia

Each teacher at High Hill Academy has been issued with a laptop computer. Natalia, an ICT teacher, uses her laptop to prepare resources, research new topics and contact her pupils using email.

When at work at her desk, she can attach her laptop to the school's network and access all the available software. She has space on one of the network's file servers to store her data files. Certain software, such as a word processor, is stored on her laptop's hard disk.

Every summer, Natalia hands in her laptop to the ICT support team for re-imaging. This procedure adds up-to-date versions of software, for which the school holds appropriate licences, and removes everything else from the laptop's hard disk drive.

1. What are the advantages to the school and to Natalia of issuing her with a laptop computer?

2. What actions should Natalia take before handing her laptop to the ICT team?

3. Why is important to carry out re-imaging on a regular basis in such circumstances?

Communication methods ◀

▶ Electronic Data Interchange (EDI)

EDI is a means of transferring information such as invitations to tender, letters, orders and invoices electronically. It allows the computers in one organisation to "talk" to the computers in their supplier's organisation, regardless of computer manufacturer or software type.

EDI cuts down the paper mountain. Although all large organisations and most smaller ones use computers, it is true that the vast majority still use considerable amounts of paper. For example, a simple order is raised on one computer, printed, mailed, then received by the supplier who re-keys the details into another computer. The process is expensive, time-consuming and prone to postal delays and errors. EDI changes all that. It collects the orders directly from one company's computer and sends them to the supplier's computer. It cuts out printed mailings, removes re-keying, minimises the margin for error and saves days in the processing cycle.

case study 3
▶ EDI at Nissan

Nissan's Sunderland plant started production in 1986. Rapid increases in production levels meant that the paperwork generated soon reached large quantities with as many as 15,000 delivery notes each week.

The labour costs of dealing with all this paperwork and the associated mailing costs were excessive. There was also the potential for human error. Following an investigation, Nissan decided to use EDI.

Almost immediately there were savings in labour and mailing costs, a shortening of the time for delivery information to reach suppliers and a reduction in the level of human errors. EDI is used to transmit delivery requirements to Nissan's logistics partner, Ryder Distribution Services, which in turn uses EDI to send delivery data to Nissan. The volume of mail to suppliers was reduced by 90–95 per cent.

1. Explain in your own words the term "EDI".

2. Find three further examples of the use of EDI.

▶ Electronic mail (Email)

With email software it is very easy to send and receive electronic messages to or from any other person or organisation that has an email address, known as a mailbox, anywhere in the world for the same cost as accessing the Internet. Most businesses now advertise their email addresses. An email address is usually of the form:

sally.miggins@computerland.co.uk

Addresses are in lower-case letters. Words are separated by full stops. No spaces are allowed. The UK at the end is the only indication of the geographical location of this address.

Email is almost instantaneous; it provides a very efficient means of communication. It saves time and the cost of postage. If the recipient of an email is not at their desk, the email is stored until they are ready to read it. An email can be forwarded to a third person; a user can send the same message to many people just by listing all their email addresses.

Internal email is suitable for memos within a business using an internal network. Email does not have the facility for a one-to-one interaction as in a telephone conversation. Overuse of email can lead to a lack of social interaction. In some organisations, it is not unusual for employees sitting at nearby desks to communicate by email rather than by talking. Generally, an email message is written in a more informal manner than a memo or a letter. There can be a tendency to abruptness which can cause misunderstanding in a way that would not occur in a face-to-face conversation.

However there are times when sending an email has a substantial advantage – an email that has been sent and replied to contains a good record of the exchange. This is not the case for a telephone or face-to-face conversation.

An email can include attachments – files that are sent with the email. For example, it is possible to send word-processed documents and images as attachments. The person receiving the email can then store and use these files in the normal way. There is a danger that an attachment from an unreliable source could have a virus that transfers to the recipient's computer when the attachment is opened.

Many people complain that they receive too many emails and consequently have to spend a considerable amount of time reading and following up on them. Some users use the carbon copy facility (cc) to send an email to many people who are not directly concerned with the message, thus clogging up the network with mail messages and increasing the amount of unwanted mail for others. Some employees waste considerable amounts of working time sending and reading personal emails; junk email (spam) can also fill up a user's mailbox.

Emails for a home user are stored on the Internet Service Provider's computer for the recipient whether or not the recipient's own computer is switched on at the time it is sent. Users have to check their mailbox to see if they have any mail. If they forget to check, email isn't very quick! Email can be sent within an organisation on a local area network.

Activity 2

Email software, such as Microsoft Outlook Express, enables users to:

- click on a Reply icon to reply to an email without having to type in the recipient's email address
- create a carbon copy of the email to be sent to a third person
- forward an email to another email address without retyping it
- set up an address book of email addresses so that the user does not have to type the email address in full every time it is used
- set up a group of several email users to whom the same email can be sent
- set the priority for an email
- store all emails sent and received
- attach files to be sent with an email.

Using email software, carry out each of the operations listed above.

case study 4
▶ **group email**

Charlie Moffat is the secretary of his local civic society. The society's executive committee meets four times a year. The committee has ten members. Whenever there was a meeting, Charlie used to type out an agenda, photocopy it ten times, place each copy of the agenda in an envelope, write a name and address on each envelope, stick on a stamp and post the letters.

Now Charlie sends out the agendas by email. He has set up a group of all the committee members in his email address book.

When Charlie writes to all the members of the committee, he creates a new email. Then he only has to select the group name in his address book and the email is sent to all ten people.

1. Describe five other features of email.

2. Summarise the benefits and limitations to an organisation of using email

▶ Fax

Fax is an alternative communications method. Like email it arrives quickly, but it gives a hard copy.

Fax is short for facsimile transmission. A fax machine uses telephone lines to transmit copies of an original document. The fax machine scans the document, encodes the contents and transmits them to another fax machine that decodes and prints a copy of the original.

The document is sent as a graphic and therefore takes longer to send than a text file. Obviously you can only send faxes to someone with a fax machine but fax machines are extremely common in business.

Unlike email which can only be received when the Internet connection is open, faxes arrive automatically. However, if information sent by fax has to be entered into a computer, it has to be retyped, wasting time and introducing the possibility of errors.

In recent years, there has been a convergence of technology in the ICT industry. Computers can be used as televisions, video and DVD players, and fax machines. Electronic fax lets you send and receive faxes from your email account. Faxes can be forwarded as email attachments.

▶ Teletext

Teletext is an electronic information service which can be viewed on most televisions. The user has to pay a little extra for a teletext TV. Teletext is operated using the TV's remote control. Teletext can be used to view information such as news, weather forecasts, TV schedules, traffic information and sports results provided by television companies. Teletext is cheap and easy to use. 80 per cent of UK households have a teletext TV, with an average of 20 million people using the service each week. Each page is numbered and transmitted in sequence. When a particular page is requested, the viewer must wait until the requested page is next transmitted.

Teletext services are offered by the BBC, ITV, Channel 4 and Channel 5. Viewers can access over 3,200 pages, with around 50,000 daily updates.

However, pages can be slow to load and the format of the page is restrictive. Few colours are used and the graphics are poor.

▶ Videoconferencing

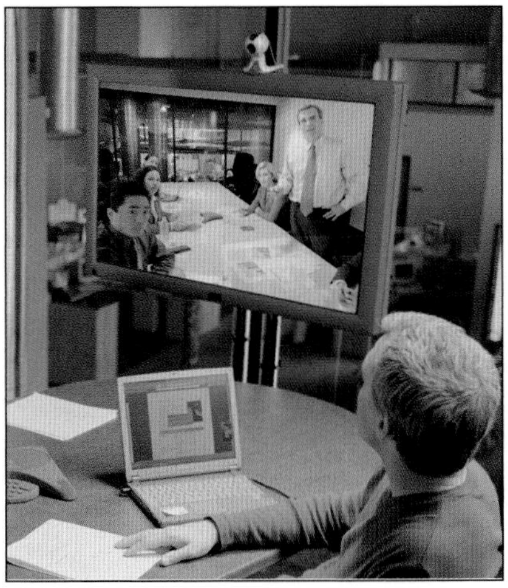

Videoconferencing, sometimes called teleconferencing, means being able to see and interact with people who are geographically apart. Two or more people can be connected to each other.

To be able to use videoconferencing facilities, a webcam, usually positioned near the monitor, and a headset consisting of a microphone and earphones are needed. A fast Internet connection and videoconferencing software are required to transmit the video images.

The home user is able to use videoconferencing at very little cost using the equipment described above with VoIP software (see Chapter 17).

In a business situation, it is also possible to use dedicated videoconferencing equipment. These machines connect directly to each other using telecommunications links.

The use of videoconferencing enables business meetings and interviews to take place avoiding the expense and time of travel. When an employee has to travel a great distance, the time taken for the journey is time that is mainly unavailable for working. Without the need for participants to travel, a videoconference can be set up at short notice. The equipment required for videoconferencing can be expensive and at present the image quality is still not as good as on television or video but it continues to develop as the hardware improves. Of course, the participants in a videoconference are dependent upon the technology. If the system were to crash, the meeting would come to an immediate end.

Figure 18.3 A videoconferencing session

case study 5
▶ videoconferencing at a national museum

Figure 18.4 Global-Leap logo

A well known national museum provides a videoconferencing service to schools throughout the UK and the rest of the world. The museum has a videoconferencing suite which is equipped with a large monitor with a camera on top. The monitor displays the participating class and has a small inset window that shows the image that is being displayed in the classroom. The suite also has built in microphones. Before a session begins, the presenter sets up a number of preset camera positions on the different objects to be discussed.

A participating school needs a videoconferencing room equipped with a large screen which can be easily seen by all pupils. Ideally, the room will have a boom mike that can be handed to each pupil who wishes to ask a question or make a comment.

About ten minutes before the start of a conference, the presenter contacts the teacher at the school to arrange for pupils to get settled in the conference room. Once connected, the presenter can explain the objects to the pupils who can be asked questions and make comments.

The set up and management of the conferencing system was organised by Global-Leap, a not-for-profit organisation funded by UK school subscriptions and donations.

1. What are the benefits of videoconferencing for schools?

2. Find out more about videoconferencing in schools at http://www.global-leap.org.

3. See other examples of videoconferencing at http://www.videoconferencing.com.

▶ **Forums**

A forum is an online discussion group. Users with common interests can exchange messages that can be read by everyone who is currently accessing the forum. Within a company, a forum can be set up for discussions between employees who can share views on important issues. It enables many people to be involved, even if they work at different locations or even in different countries.

On the Internet, a forum is often called a **newsgroup**. There are thousands of newsgroups covering every interest you can think of. To view and post messages to a newsgroup, you need a news reader, a program that a user runs on his computer. It connects to a news server on the Internet. The Web browsers Microsoft Internet Explorer and Netscape Navigator both include a news reader.

Discussion forums are widely used in education. The Open University, where students study from home, provides discussion forums relating to many of its courses. Students who are studying alone at home may have very little direct contact with their fellow students. The use of a discussion forum provides them with a chance to reduce any feeling of isolation by sharing ideas and problems. It can allow them to develop a greater understanding of the subject being studied.

▶ Worked exam question

A multi-national company has offices in London and New York.

1. How can the company use ICT to enable effective and efficient communication? (9)

2. There can be problems with communicating using ICT. Explain **three** disadvantages of using ICT for communication. (6)

▶ **EXAMINER'S GUIDANCE**

As there are nine marks available for your answer to part 1, you need to identify three ways in which ICT could be used to aid communication. One mark is available for each method you identify and a further two marks for explaining why each method is of benefit. You must ensure that you come up with two distinct reasons. The most obvious methods to discuss are videoconferencing, the use of email and discussion forums. Take care that you don't give the same reason for different features – for example you would get a mark for writing that videoconferencing allows colleagues at different locations to communicate. You could not then get a mark for saying that a forum allows colleague at different locations to communicate as well.

▶ **SAMPLE ANSWER**

Videoconferencing can be used for meetings (1) as there is no wasted time when employees are travelling (1). The employees can arrange a meeting at short notice and the company can save the cost of travel (1).

Email can be used to communicate with colleagues (1). This provides efficient communication as the email is received almost instantly (1). It also saves the cost of postage (1).

Forums can be set up for discussions (1); employees can share views on important issues within the company enabling all to be involved (1) wherever they are located (1).

▶ **EXAMINER'S GUIDANCE**

In part 2, again you need to think of three reasons but they must be different. When checking your answer at the end of the exam, make sure that the number of points that you have written is the same as the question asked for. Students tend to write at length about one point and forget how many points (in this case, disadvantages) they were asked for.

▶ **SAMPLE ANSWER**

The lack of face-to-face contact when using email (1) may result in a discussion that may not be as effective as meeting in the same room (1).

When a group of people use videoconferencing, the system may crash (1) so the meeting would come to an immediate end (1).

Email communication can sometimes be abrupt (1) causing a misunderstanding between employees (1).

ICT networks for different geographic scales and uses

A network may be restricted to one room or one building or cover a small geographical area. Such a network is called a **local area network** (LAN). A LAN was traditionally connected via direct lines – physical links using its own dedicated cables. These can be **twisted wire**, **coaxial** or **fibre-optic** cable. The development of wireless networks means that many small LANs now operate without physical cables. Small, home networks where several computers are linked using a wireless router were discussed earlier. Many small businesses use a peer-to-peer local area network where resources can be shared very easily.

---

case study 6
▶ **a graphic design studio**

Three graphic designers, Adrian, Mike and Sanjit, have decided to set up a new business together. They will employ two assistants. They have to decide on the IT equipment they need to buy for the business. One of the decisions they need to make is whether to install a small network or to have stand-alone computers.

If each designer were to have his own stand-alone computer, they would each need specialist peripherals: a scanner and a high-quality printer. This would prove expensive. However, there would be no need for one of them to wait while a device was in use by someone else, as might occur with a network.

If they chose to install a network, the designers and assistants would be able to share data; this would allow two or more of the team to work on the same project.

A network would allow the team to communicate with each other through email, share online diaries and access each other's documents for proofreading.

1. The team have to decide whether to install a peer-to-peer or a server-based network. Discuss the issues involved.

2. Discuss the benefits and disadvantages to the designers of having Internet access.

A network may be spread over a wide geographical area, possibly covering different countries. Such a network is called a **wide area network (WAN)**. It can be linked by public telecommunications systems such as telephone lines, satellite links and microwave signals.

The **Internet** is a very large WAN, so a home user who accesses the Internet is using a WAN.

Many larger organisations have their own server-based local area network linked into a wide area network.

case study 7
▶ the National Lottery

Figure 18.5 Lottery ticket

The UK National Lottery, run by Camelot, sells tickets in around 35,000 retail outlets. Tills in all the retailers are connected to Camelot's wide area network either by cable or by satellite. As lottery tickets are sold, details of the numbers chosen are entered by optical mark reading (OMR).

The data is transmitted to Camelot's computer centre in Rickmansworth, Hertfordshire. The network needs to be very sophisticated to cope with the large volumes of sales (particularly early in the evening before the draw is made) which have reached over 50,000 transactions a minute. Camelot say that the network has been designed to cope with considerably more traffic than this.

1. What data needs to be transmitted from a lottery network station to the central computer?

2. Discuss the reasons why Camelot use their own WAN for collecting data from shops rather than linking into the Internet.

3. Describe the hardware that would be needed to enable lottery sales in a new store.

case study 8
▶ networking in an FE college

Faxton College is based on two sites, a mile and a half apart. All the students on the main site are full-time students aged 16–19; the second site is used for part-time adult courses. All financial, examination support and personnel functions are carried out at the main site. Each site has had its own managed LAN for several years and both have been managed by the shared IT team. Software had to be installed and maintained on two servers, one for each site. Computers were connected by cable.

A few years ago, the IT Manager decided to link the two LANs with a WAN. The link between the two LANs was established using a dedicated telephone line. Central file servers on one site hold the database file management information software for the whole college, and staff on both sites share the software and access common information. This means that the IT support team only have to maintain the software in one place; when upgrades are issued, they only need to be applied once. Members of staff who work on both sites can easily access all their files stored on their personal network drive. One further benefit of installing a WAN means that files can be backed up in a different location.

1. What is the difference between a LAN and a WAN?

2. Why is the ability to back up files on a separate site seen to be an advantage?

3. If you were wishing to link the two LANs today, in what other ways could you provide the link?

Very extensive networks exist which are often called **metropolitan area networks (MAN)**. JANET is an educational wide area network that connects UK universities, FE colleges and other educational establishments. There are over 18 million users of the JANET network. It also provides Internet access for users.

JANET supports a wide range of facilities including videoconferencing and video streaming, which allow lectures to be delivered to remote groups of students.

Large international companies such as IBM and Unilever maintain their own wide area networks which are independent of the Internet. They have their own secure links between sites. This means that sensitive data is not transferred over the public network and is therefore easier to keep secure.

Protocols and standards ◄

► Standards

If you bought a new computer system, you would expect to be able to transfer data from your old machine. If you bought a more recent software version, you would expect it to be able to read data from the previous version. If you bought a digital camera, you would expect to be able to transfer the images to your PC. You can expect these things because standards are in place.

Hardware standards

Today most computers have a number of **USB (Universal Serial Bus)** ports, which are used to connect a wide range of peripherals such as digital cameras, hard disks, modems, scanners, printers and mice. This equipment must conform to standards, which means that buyers are no longer restricted to one company. Competition means cheaper prices for the buyer.

Software standards

The use of software standards should make applications easier to use by having a common feel. It is standard that the F1 key is used for Help. The names of menus and their positions are consistent. Software can usually save work in a variety of different formats which enable portability and compatibility. Information can be viewed on the Web if it is saved as an HTML file. Most modern packages have this feature.

► Protocols

A protocol is a set of formal rules and procedures that define how devices can communicate. Without protocols there would be no agreed way in which a computer could transfer data to and from another computer.

Protocols enable the use of open systems – computer systems that can communicate regardless of the manufacturer and the platform. This is very important for the Internet, which can be accessed by a wide variety of hardware devices such as digital TV sets, mobile phones and PDAs.

The Internet uses a number of internationally-agreed standards and protocols which mean that it can be accessed by a variety of hardware platforms.

The standards ensure that there is a reliable connection between devices and provide error detection and correction mechanisms. The following protocols are used on the Internet:

- **File Transfer Protocol (FTP)** allows a file to be transferred from one computer to another; it is often used to upload files to the Web.
- **HyperText Transfer Protocol (HTTP)** is a standard for transferring Web pages to a client computer.
- Post Office Protocol (POP) is a standard for transferring email between computers.
- **Transmission Control Protocol/Internet Protocol (TCP/IP)** allows Internet users and providers to communicate with each other, no matter what hardware is used.
- **Wireless Application Protocol (WAP)** is a standard for wireless communication networks used by mobile phones to access the Internet.

▶ Benefits and limitations of standards

The main advantage of standards is that the user is not restricted to one manufacturer's equipment. Even if one company's computers are all the same make, they may wish to communicate with another company whose hardware is different, for example, to use EDI. This would not be possible without protocols. Being able to choose from a variety of manufacturers means there is competition and prices are likely to be lower.

The main disadvantage of standards and protocols is that they are difficult to change. It is difficult and takes time to get universal agreement on the establishment of new standards. In a fast-changing area, standards may not be able to keep up with technological developments. Open systems based on old standards may be unacceptably slow.

Having to follow standards means that in some cases the full power of the machine might not be available and there might therefore be reduced functionality or performance. Bespoke software, designed specifically for use on a particular platform and ignoring standards, makes better use of the hardware.

▶ Development of protocols and standards

De jure standards

Many standards are formally introduced, often after considerable deliberation by an international committee. These are called *de jure* standards and include:

- ASCII codes were devised by the American Standard Code for Information Interchange.
- The JPG image format was developed by the Joint Photographic Experts Group (JPEG), which represents a wide variety of companies and academic institutions worldwide.
- Most Internet protocols, such as TCP/IP, were drawn up by CERN (the European Organisation for Nuclear Research, an international group of 20 member countries based in Geneva, Switzerland).

De facto standards

Other standards, known as *de facto* standards, arise through historic precedence or as a result of the marketing and sales success of a particular product. Examples of *de facto* standards include:

- MS-DOS and Microsoft Windows have become the standard operating system and GUI for PCs.
- The GIF image format was created in 1987 by Internet company CompuServe as a format for transmitting images over the Internet.
- The standard 3½-inch floppy disk was introduced by Sony in 1980. There were many competing formats but, over time, the industry settled on the 3½-inch format.
- USB 1.1 was developed in 1998, by companies that included DEC, IBM, Intel, Microsoft, and Compaq. USB 1.1 was integrated into Microsoft Windows 98.
- PDF format is used for document sharing on the web. PDF is an open standard that was developed by Adobe.
- RSS is used for the sharing of Web content. Its most widespread use is in distributing news headlines on the Web.

Often *de facto* standards have evolved not because they are technically the best but due to commercial or other pressures. In the 1980s, there were two types of video recorder: VHS and Betamax. Betamax was widely regarded as being better quality but VHS became more popular due to better marketing. Betamax flopped while VHS became the standard.

A similar situation is happening in ICT. Microsoft MS-DOS and Windows have such a dominant market share that they have become the standard operating system for a PC. It doesn't mean they are the best. Many people swear by Mac OSX or Linux. However, as Windows has such a large market share, software developers are more likely to be interested in producing new software for Windows than, say, the Linux operating system.

case study 9
▶ Wi-Fi standards

802.11 is a standard used for wireless local area networks (WLANs). It was developed by the Institute of Electrical and Electronics Engineers (IEEE), an international organisation that develops standards for hundreds of electronic and electrical technologies.

Networks using the 802.11 standard are often called Wi-Fi, short for wireless fidelity. As technology has improved, different versions of the standard have been developed using different radio frequencies and offering different speeds of data transfer.

802.11b is a standard for WLANs offering speeds of up to 11 Mbps; 802.11g is a new standard offering speeds of up to 54 Mbps. The latest Wi-Fi standard (802.11n) will bring faster transmission speeds of up to 100 Mbps.

What users will want to know is if increased performance justifies increased cost and if there will be problems on a mixed network.

1. What is a standard?

2. Is Wi-Fi a *de facto* or a *de jure* standard?

3. Why is a standard such as 802.11 necessary?

4. What is the main disadvantage of standards such as 802.11?

5. Use the Internet to research the latest 802.11 standards for WLANs.

SUMMARY

Home use of ICT includes:

▶ keeping in contact with friends and family using email and VoIP

▶ accessing the Web for information and services

▶ online shopping and banking

▶ downloading, viewing, modifying and sharing photographic images

▶ accessing social networking sites (SNS)

▶ maintaining a Web log

▶ instant messaging.

Connecting to the Internet from home can be achieved using:

▶ a dial-up facility

▶ broadband using ADSL technology

▶ a cable modem, in many towns and cities

▶ wireless broadband, available in certain areas of the country.

A variety of methods can be used for communicating electronically:

▶ Electronic Data Interchange (EDI) allows businesses to exchange data electronically.

▶ Email enables users anywhere in the world to communicate quickly and cheaply.

▶ Fax allows a hard copy of a document to be transmitted via a telephone line to anywhere that has suitable hardware.

▶ Teletext is an electronic information service which can be viewed on most televisions.

▶ Videoconferencing is a method of allowing two or more people in different locations to interact. A webcam and microphone are needed at each location.

▶ A forum is an online discussion group.

A local area network (LAN) is a network that covers a small geographical area. A wide area network (WAN) is a network that is spread over a wide geographical area.

A protocol is a set of formal rules and procedures that define how devices can communicate. The Internet uses a number of internationally-agreed standards and protocols. *De jure* standards are formally introduced by an international committee. *De facto* standards arise through historic precedence or as a result of the marketing and sales success of a particular product.

Questions

1. Explain the difference between a LAN and a WAN. (2)

2. Many households in the UK have access to the Internet.
   a) Describe the different ways that a computer can be linked to the Internet. (6)
   b) Using examples, describe different uses of the Internet and how these uses could benefit individuals. (14)

3. A general-practice team of doctors, nurse, practice manager and receptionist currently uses a number of stand-alone computers to manage patient records, appointments, correspondence and all financial accounts. The practice manager is considering installing a network.
   a) Describe the advantages to the practice of installing a network. (6)
   b) Discuss the type of network that would be appropriate. (4)

4. Discuss the benefits and limitations of communicating using email. (6)

5. A worldwide company is installing videoconferencing facilities in its offices.
   a) Describe the hardware that is required for videoconferencing. (4)
   b) Describe the benefits to the company of installing videoconferencing facilities. (6)

19 *Protecting data in ICT systems*
AQA Unit 2 Section 5 (part 1)

Privacy of data in ICT systems ◀

The development of information technology has meant that many organisations store details about you electronically on an ICT system, for example:

- Your school or college stores details of your courses, exam results and contact details.
- Your doctor stores details relating to your health.
- Your bank stores details of your financial transactions.
- If you have a part-time job, your employer stores your employment details.
- If you have a driving licence, your details are stored at the DVLA.
- Stores may have details about you if you have a reward card or have bought products online.
- Your local council stores details of everyone who is 17 or over so that it can produce the electoral register.

These are all examples of personal data – data about living people.

Activity 1

Make a list of all the organisations that you think store information about **you**.

The increase of personal data stored on computer has worried many people. Their main concerns are:

- Who will be able to access this data? There is a fear that personal data could be accessed by unauthorised people who could use it to defraud an individual. Will information about me be available remotely over a network and therefore vulnerable to being accessed, resulting in identity theft for example? Could my medical records be examined by a potential employer?
- Is the data accurate? If it is stored, processed and transmitted by computer, who will check that it is accurate? People often think it must be true if "it says so on the computer". Inaccurate personal data that is stored could have an adverse effect on an individual. For example, if

inaccurate data is stored regarding payment of bills, an individual might be refused a credit card or a loan.

- Will the data be sold on to another company? For example, could my health records be sold to a company where I have applied for a job? Can my school records be sold on to someone else? Could my personal details, collected by my employer, be used by a commercial company for targeting junk mail?
- How long will data be kept? As it is very easy to store vast amounts of data, will data about me be stored even if it is not needed? For example, if I apply for a job but don't get it will the data be deleted?

What is personal data? ◀

Personal data covers both facts and opinions about a living person. Facts are pieces of information, such as name, address, date of birth, marital status or current bank balance. Results in examinations, details of driving offences, a record of medications prescribed and financial credit rating are further examples of facts that could relate to an individual. Personal opinions such as political or religious views are also deemed to be personal data.

Think before you give away personal information, you never know where it will end up!
(Advertising campaign slogan)

Data protection legislation ◀

The concerns mentioned above about the use of personal data led to the Data Protection Act 1984. The 1984 Act is a law that set outs regulations for storing personal data that is automatically processed.

The Data Protection Act 1998 strengthened the 1984 Act and enshrined the European Union directive on data protection into UK law. This means that UK law is in line with the data protection laws in all the other countries in the European Union.

▶ What the Data Protection Act 1998 says

The Data Protection Act 1998 sets rules for the electronic processing of personal information. The law also applies to paper records from 23 October 2007.

The law refers to:

- **data subjects** – people whose personal data is being processed
- **data controllers** – people or organisations who process personal data.

  The law works in two ways:

- data subjects have certain rights
- data controllers must follow good information-handling practices.

▶ What data controllers must do

Data controllers must follow the eight data protection principles. They must register the fact that they are storing personal data with a government official called the **Information Commissioner**. The following information must be registered:

- the data controller's name and address
- a description of the data being processed
- the purpose for which the information will be used
- from whom the information was obtained
- to whom the information will be disclosed and countries to which the data may be transferred.

Activity 2

It is possible to use the Internet to find out whether a company has registered. Go to **http://forms.informationcommissioner.gov.uk/search.html**.

▶ Data protection principles

The data protection principles say that data must be:

1. fairly and lawfully processed
2. processed for registered purposes
3. adequate, relevant and not excessive
4. accurate and up to date
5. not kept for longer than is necessary
6. processed in line with your rights
7. secure
8. not transferred to countries without adequate protection.

Principle 1 means that you cannot collect data for one purpose and then use it for another purpose (even if the purpose is registered) without the permission of the data subject.

Principle 2 means that if a company intends to sell data on to another company it must register this with the Information Commissioner. (They will need the permission of the data subjects to do this.)

Principle 3 means that any irrelevant data should be deleted. For example, data about unsuccessful job applicants should not be kept.

Principle 4 means that the organisation must take steps to ensure that its data is accurate. Once a year, a school, for example, may provide each pupil (the data subject) with a printout of their personal details for checking purposes.

Principle 6 means that data subjects have the right to inspect the data held on them, for payment of a small fee. They have the right to require that inaccurate data is corrected. They have the right to compensation for any distress caused if the Act has been broken.

Principle 7 means that appropriate technological security measures must be taken to prevent unauthorised access. This means information has to be kept safe from hackers and employees who don't have the right to see it. Your data can only be passed on to someone else with your permission. Backup copies should be taken so that data is also safeguarded against accidental loss.

When you fill in the form in Figure 19.1 to register with the Guardian Unlimited website, your details can be passed on if you tick the last box.

Principle 8 means that personal data cannot be transferred to countries outside the European Union unless the country provides an adequate level of protection.

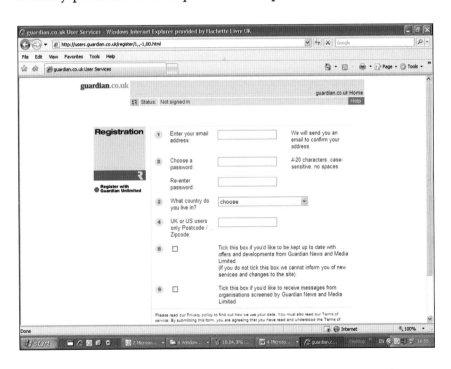

Figure 19.1 Online registration form for the Guardian Online

P – registered **purposes** (2)

E – not transferred outside the **EU** (8)

R – **relevant** (3)

S – **secure** (7)

O – you can inspect your **own** records (6)

N – **not kept** for longer than is necessary (5)

A – **accurate** (4)

L – **lawfully** processed (1)

Activity 3

Test yourself. Close the book and try to name five of the principles.

▶ Is your consent needed to process your data?

Suppose you apply for a supermarket loyalty card. The supermarket needs to store and process your personal details as part of their normal work. They do not need your consent to do this – you have agreed to this when you applied for the card.

However, the supermarket cannot pass your details on to another company without your consent. In practice, this means that when you fill in the application form, you can tick a box to prevent your personal details being sold on to another company, for example, for direct marketing.

Note: If you cancel the card, your details are no longer needed and should be deleted by the supermarket. At any time you can write to an organisation that is sending you junk mail asking them to stop processing any data about you.

▶ Exemptions from the Act

Information is exempt from the principles of the Data Protection Act if it is used:

- to safeguard national security
- to prevent and detect crime
- to collect taxes.

Personal data relating to someone's family or household affairs does not need to be registered.

▶ What is the role of the Information Commissioner?

The Commissioner has the responsibility of ensuring that the data protection legislation is enforced.

The Information Commissioner keeps a public register of data controllers. Each register entry must include the name and address of the data controller as well as a description of the processing of personal data carried out under the control of the data controller. An individual can consult the register to find out what processing of personal data is being carried out by a particular data controller.

The Data Protection Act 1998 requires every data controller who is processing personal data to notify the commission unless one of the exemptions listed in the Act applies. At the commission office, a complete copy of the public register is kept and it is updated weekly.

Other duties of the Commissioner include promoting good information handling. As well as keeping the register of data controllers, the Information Commissioner also gives advice on data protection issues, promotes good information handling practice and encourages data controllers to develop suitable codes of practice. He also acts as an Ombudsman.

Note: The Information Commissioner was called the Data Protection Registrar by the 1984 Act.

---

case study 1
▶ lack of security

Danny Hughes is an A-level student who had a holiday job on the production line at Betta Biscuit plc. One lunchtime, Danny decided to explore the factory and found his way into the computer room. There was no one about. Danny sat down at a terminal and typed in a few usernames with no luck. Then he noticed a birthday card on the desk: to Bob with love from Jane. Danny typed in the username BOB and was asked for a password. He typed in JANE and, to his surprise, it was accepted!

A menu appeared on the screen. Danny chose payroll. He could load up the payroll information of all the employees. Danny loaded the file of his friend Chris and cut his hourly pay by half. Two workers came in and saw Danny, but no one said anything. Danny logged off quietly and slipped out of the room undetected.

1. Suggest as many steps as you can that Betta Biscuit plc should take to improve their security.

2. Who has broken the law, Danny or Betta Biscuit or both of them?

case study 2
▶ is it legal?

A company registers the following use with the Information Commissioner:

■ Purpose: The administration of prospective, current and past employees

■ Typical activities: Payment of wages, salaries, pensions and other benefits; training; assessment and career planning

■ Type of information: Names, addresses, salary details and other information related to their work

The company wants to sell the names and addresses of their employees to another company for direct marketing. This is illegal; it is not the registered purpose so it breaches Principle 2.

The police want the company to give them information about an employee in relation to a possible fraud. This would be legal as long as the police provided a certificate of exemption that this was connected with the detection of crime.

Copy the table below and fill in the blank cells for a charity registering with the Information Commissioner.

Purpose:	The administration of donations
Typical activities:	
Type of information:	

Activity 4

You can find out more about the Data Protection Act and the Information Commissioner at http://www.dataprotection.gov.uk/. Explore the site and answer the following questions:

1. What are your rights in preventing junk mail?
2. What two laws does the Information Commissioner oversee and ensure compliance with?

case study 3
▶ DPA in the news

Marks and Spencer admit breaking DPA for 15 years
In 1999, the high-street chain store Marks and Spencer had to change its procedures after learning it had been breaking the Data Protection Act for almost 15 years.

The company had been disclosing charge-card account details to supplementary card holders – people who are authorised to charge goods to another person's account.

case study 3 (contd)

Lloyds TSB accused of breaking the DPA

In 2004, a customer alleged that Lloyds TSB had broken the Data Protection Act when it was transferring work to India. The bank was accused of sending its customers' personal financial data outside the European Union without their written consent.

Organ group president is fined

In 1996, Trevor Daniels, the president of the Association of Organ Enthusiasts, was fined £50 for keeping his membership list on his home computer without being registered.

Solicitor fined for failing to notify

A solicitor, Ralph Donner, was prosecuted by the Information Commissioner's Office in 2005 for failing to notify under the Data Protection Act 1998. He was fined £3150 and ordered to pay £3500 towards prosecution costs. Mr Donner had been contacted more than five times by the Information Commissioner over a period of two years without notifying.

Complaint against mobile phone company

In 2007, a complaint was made to the Information Commissioner about the way in which a mobile phone operator processed personal information. New members of staff were allowed to share usernames and passwords when accessing the company IT system.

Utility companies break DPA

In 1997, the Information Commissioner reprimanded two utility companies for breaking the Data Protection Act. The companies stored names and addresses for sending bills to customers. They then used these details to send out direct mail advertisements to their customers.

1. Which data protection principle did Marks and Spencer break?

2. Which data protection principle was Lloyds TSB accused of breaking?

3. What did Mr Daniels do wrong?

4. What should Mr Donner have done?

5. Which data protection principle did the mobile phone company break?

6. Which data protection principle did the utility companies break?

7. Use the Internet to search for details of other breaches of the Data Protection Act. Use sites such as http://www.guardian.co.uk/ or http://www.independent.co.uk.

It is important that you are aware of when you are giving permission for your personal data to be used. You need to read the small print of any conditions before signing up to an agreement. Consider carefully whether or not you wish to allow your personal data to be handed on to others before you agree to allow the company to do so.

Data stored on portable storage media should be looked after very carefully and kept safe.

Another way in which personal data is acquired by third parties is through the carelessness of a user. When a user disposes of an old computer they often fail to remove all the data that is stored on the hard disk. This could include personal finance data, address lists of friends and contacts, photographs and personal correspondence.

To protect against such unauthorised access to computer material you should delete all the data from the hard drive (see Case Study 4).

case study 4

► **hospital disk raises fears about protecting personal data**

(This is from an article by Pete Warren of The Guardian in September 2007.)

For the past month or so, Dudley Group of Hospitals NHS Trust has been dealing with a problem that should not have happened – all because a computer hard drive containing sensitive patient information from a trust hospital was sold on the auction site eBay. The disk contained data on cancer patients at the hospital.

And Dudley is not alone. One hundred and thirty three discarded disks obtained in the UK were analysed and 75 were found still to be working. Data was found on 62% of those – including company records, personal information, financial data and paedophile material which resulted in a police investigation. With some of the records, the information was so detailed that the individuals could have had their identities stolen.

The destruction of data about individuals by companies is a legal requirement in disposing of a computer. But just deleting a file or reformatting the drive does not remove the data; it removes the file entry from the index, marking the space as available for reuse, and the data can be restored. To prevent that, overwrite the data. Encrypting a disk is a good first measure: Windows XP and Vista, Mac OSX and Linux all offer this facility. Then get a disk-wiping program: there's a list including free ones and instructions at http://howtowipeyourdrive.com. Some erase the entire disk while others can select files or folders to erase. Ensure you erase free space.

1. Why is it a legal requirement to destroy data when a computer is disposed of?

2. Explain the relevance of the Data Protection Act to the cases described in the case study.

Intrinsic value of data

It is hard to put a monetary value on most data that is stored in an ICT system. However, this does not mean that the data has no value to the organisation. The availability and flow of data will affect the performance of a company. In extreme cases, the loss or corruption of data could cause the business to fail.

Data has an intrinsic value – a value in its own right. It is easiest to understand this by looking at the consequences that arise if data is lost.

Consider an airline's flight booking system. Data is stored on all flights that have already been booked. If some of the data were to be lost then all trace of a customer's booking could be wiped out. The seats that were booked could be booked by someone else at a later time; when the original customer arrives at the airport to check in there would be no trace of the booking. This would cause extreme customer dissatisfaction and if such a loss occurred on a regular basis, the airline's reputation would be damaged severely.

Commercial value of data

Data can have a financial value. In many ways, it is a commodity like oil, gold or wheat.

An organisation may build up a database consisting of names and addresses of customers or contacts that would be valuable to another organisation.

For example, a charity might have a list of names and addresses of donors that another organisation could use to target individuals by mailing letters to advertise an event likely to be of interest to the donors. Such a targeted mailing would be more effective than sending out letters to everyone living in the area as the letters sent in the targeted mailing would only go to people who are likely to be interested in the event. Money would be saved as fewer letters would need to be sent.

The organisation could collect the name and address data for itself, rather than purchase it from the charity, perhaps by undertaking a survey. However, this method of obtaining the data would be more expensive and time consuming than purchasing it from the charity. The data only has a monetary value for a time, whilst it is sufficiently up to date to ensure that most names and addresses are still current. As time passes, many people may have moved house or their circumstances or interests will change. More letters are wasted (and therefore more money) and eventually the time and money spent on the mail shot is no longer worthwhile.

Whether or not data has monetary value to an organisation will depend upon the potential use to which it can be put.

▶ Obtaining data

Whenever goods are purchased by telephone, mail order or online via the Internet, data about the customer is gathered; whenever a form requesting a "special offer" for goods or information is sent to a newspaper or magazine, whenever a person enters a competition, the consumer's details are likely to be stored electronically.

Organisations have a legal requirement to ask customers if their data can be passed on to others. This is often done by using a tick box with a message such as: "Information gathered in this way may be used to target specific customers through direct mail."

▶ Free data

Much data is freely available to the public. The electoral register is compiled by local councils and lists the names and addresses of people entitled to vote in elections. A section of the electoral register might look like this:

Paddock View, Pillsbury, P85 3RH

1 Paddock View	**Albert Mitchell**
1 Paddock View	**Doris Mitchell**
2 Paddock View	**Sally Bryant**
3 Paddock View	**Graham Williams**
3 Paddock View	**Sally Williams**
3 Paddock View	**Daniel Williams**
3 Paddock View	**Adam Williams (10 August)**
4 Paddock View	**William Hunter**
4 Paddock View	**Shirley Hunter**

There are a number of things that can be deduced from this information which businesses might find valuable. For example, all nine people live in Pillsbury. A local restaurant might see them as potential customers. A date appears after Adam Williams' name. This means that he will be old enough to vote on that day and so is approaching 18. He might be a potential customer for the local nightclub.

Data on recent births, engagements, marriages and deaths appear in local newspapers. Local government departments give information on planning applications for buildings.

Data is valuable and there are costs involved in collecting it. Even when data is available free, for example, the electoral register, there are labour costs in entering the data and converting it into a suitable electronic format. The electoral register on paper is not very convenient.

Companies such as Millennium Data Ltd (http://www.marketinglists.com/eroll.htm) resell the electoral roll in electronic format. They say this electronic format is useful for:

- direct marketing campaigns
- political campaigns
- data analysis, capture and validation
- software development.

Telephone directories contain alphabetic lists of subscribers together with their telephone number. It is not practical to use such a directory to obtain the name and address of the holder of a particular number. However, when the data are stored electronically, such searches can easily be performed. There are companies that scan in telephone directories and sell customers the information in electronic format.

Data on all house purchases are stored by the land registry. A number of companies have websites that allow users, for a fee, to find the value of recent sales of houses near to them. Try the sites http://www.myhouseprice.com and http://www.ourproperty.co.uk (see Figure 19.2).

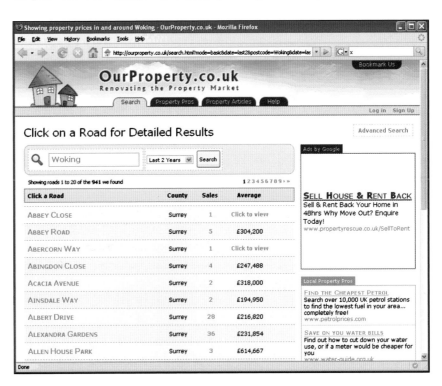

Figure 19.2 Property website

The more data that is required and the more detailed the data, the more it will cost to obtain. If the data are collected through a survey, the cost will depend on the size of the sample and the number of questions asked.

case study 5
▶ **supermarket loyalty cards**

The use of loyalty cards is common in supermarkets and other large stores. Customers have to fill in an application form to obtain a card; this requires them to give their name, address and other information. The loyalty card is used whenever a customer purchases goods and points are allocated according to the amount spent. The points can then be used to make further purchases.

The use of loyalty cards means that special offer information and vouchers can be sent to customers. The types of goods bought can be linked to a customer. The loyalty card identifies the customer. Its identity number is input into the computerised till when details of purchases have been gathered from the bar codes on the products. Thus the special offers sent to a customer can be for products that are likely to be of interest.

This data relating to the purchasing patterns of a customer is of use to other organisations besides the supermarket itself. For example, a list of names and addresses of people who regularly purchase cat food would be valuable to a company selling pet insurance products.

1. List three other types of company that would be interested in purchasing the list of names and addresses of pet owners.

2. List five types of company that would be interested in purchasing a list of names and addresses of people who regularly buy baby food. Explain why it would be important that the list should not be more than three months old.

3. Explain why it is worthwhile for the organisations listed above to buy the list of names and addresses from the supermarket.

case study 6
▶ **Porsche cars**

Porsche Cars Great Britain (PCGB) imports all Porsches into Britain and owns five Porsche dealerships. Sales are small compared with volume manufacturers such as Ford or Vauxhall.

PCGB know that the people most likely to buy one of their cars are people who have bought one in the past. These people obviously have an interest in sports cars and presumably have the necessary income to purchase one.

This means that PCGB spend 80 per cent of their marketing budget on direct mail targeting previous customers. They have built up a large database of 22,000 current Porsche owners, over half of all Porsche owners in the country. These owners receive a copy of Marque, PCGB's magazine. This is an important part of their marketing strategy. Industry sources suggest that 60 per cent of customers will buy from you again simply because you keep in touch.

case study 6 (contd)

Keeping a database like this is quite legal but PCGB must comply with the Data Protection Act.

1. Give three reasons why PCGB use this method of promoting their cars rather than advertising on television as volume car manufacturers do.

2. Give three further products, other than cars, that might best be marketed in this way.

3. Describe the measures PCGB would need to take to ensure that their database of customers is kept accurate and up to date.

SUMMARY

The Data Protection Act concerns the storage of personal data. Data controllers must:

▶ **register with the Information Commissioner**

▶ **follow the eight data protection principles.**

The principles say that personal data must be kept secure and be accurate, up to date and only used for the registered purpose.

Data subjects have various rights under the Act including the right to inspect data about themselves, have any errors corrected and claim compensation for any distress.

Personal information involving national security matters, the detection of crime and the collection of taxes is exempt from the Data Protection Act.

Data has an intrinsic value. Loss of data can have far-reaching effects on a company.

Information is a commodity that can have a monetary value. The value depends on its accuracy, potential use and its intended use.

Information that is freely available may have a monetary value when it has been converted into a more useful electronic format.

Ensuring that information is up to date can be time consuming and costly.

Information that is protected by law from disclosure can still have a commercial value.

Questions

◄

1. The Data Protection Act of 1998 refers to an "Information Commissioner" and a "data subject". Explain what is meant by these terms. (4)

2. Explain what is meant by the term "personal data". (2)

3. Explain why it is necessary to have data protection legislation. (6)

4. A company wishing to store customers' personal data must register their use. In addition to details about the company, state three items of data that a company must include in their entry on the Data Register. (3)

5. The Data Protection Act 1998 is an act designed to regulate the processing of personal data.
   a) State what is meant by "personal data". (1)
   b) State with whom a company should register if it stores personal data, (1)
   c) Explain the rights of a data subject who thinks that the data stored is incorrect. (2)

6. Explain what is meant by the intrinsic value of data. (3)

7. Discuss how data can have a commercial value. (6)

8. An optician keeps records in a database of all its customers who have had eye tests. Eye-test reminders are sent out to customers when they are due. Customers who do not make appointments after two reminders have been sent out have their details deleted from the database. Describe two possible reasons why these customer details are deleted from the database. (4)

   AQA January 2003 ICT1

9. The personnel department of a large company keeps records on all the employees of the company. These records contain personal data and details of the employees' position, training and medical history. The company is registered on the Data Protection Register and has to abide by the principles of the 1998 Data Protection Act. Three of these principles are:
   ■ Personal data shall be adequate, relevant and not excessive in relation to the purpose or purposes for which they are processed.
   ■ Personal data shall be accurate and, where necessary, kept up to date.
   ■ Appropriate technical and organisational measures shall be taken against unauthorised or unlawful processing of personal data and against accidental loss or destruction of, or damage to, personal data.
   For each of the principles stated above, describe what the company must do to comply with them. (6)

   AQA June 2002 ICT1

10. When installing double glazing with a large national company, customers are asked if they object to the data they are giving to the company being passed on to other companies.
    a) Explain why the company must ask this question. (2)
    b) Describe what the double-glazing company could do with the customers' details if they give permission for them to be passed on. (2)

20 *Protecting ICT systems*
AQA Unit 2 Section 5 (part 2)

ICT systems are vulnerable; they need to be protected to ensure that hardware, software and data are kept safe.

Internal and external threats

ICT systems are at risk from external threats which come from outside the organisation. Such threats include attack from viruses and theft of money or data when an outsider breaks in to an ICT system.

An organisation is subject to internal threats to its ICT systems from its own employees. These can be deliberate or can occur through carelessness.

▶ What is computer crime?

Computer crime is any illegal act that has been committed using a computer as the principal tool. As the role of computers in society has increased, opportunities for crime have been created that never existed before.

Computer crime can take the form of:

- the theft of money (for example, the transfer of payments to the wrong accounts)
- the theft of information (for example, from files or databases)
- the theft of goods (by their diversion to the wrong destination).

Committing a crime breaks the law as passed by Parliament. Crimes are punished through the courts. Punishments are likely to take the form of a fine or, in severe cases, prison.

Examples of computer crime include gaining unauthorised access to an ICT system, illegally copying a piece of software from a computer in the workplace and taking it home or applying for a loan via the Internet using a false identity.

▶ What is malpractice?

Malpractice is defined as negligent or improper professional behaviour. It is the act of breaking professional rules set by the employer or professional body that results, intentionally or unintentionally, in causing harm to their organisation or clients. An employee who carelessly leaves his workstation logged on, or divulges his password to others, could be enabling unauthorised access to data. This could be considered to be malpractice.

Excessive use of a computer at work for personal use by an employee could be considered to be malpractice. With so many workers having a computer with Internet access on their desks, there is a growing concern about Internet misuse at work. This could be through: wasting time surfing the net; e-shopping; sending and receiving personal emails; accessing inappropriate sites, such as those displaying pornographic images, or visiting chat rooms.

Punishment for malpractice is likely to be a warning, downgrading, dismissal or expulsion from a professional body depending on the severity of the malpractice.

---

case study 1
▶ 28 arrested in global web fraud sting

(From Robert Jaques, http://vnunet.com, 29 October 2004)

Police have arrested 28 individuals suspected of being part of a global Internet-based organised crime network dealing in identity theft, computer fraud, credit card fraud and conspiracy.

A 19-year-old British man from Camberley, Surrey, was arrested as part of the sting operation on Wednesday by National Hi-Tech Crime Unit (NHTCU) officers.

Led by the US Secret Service, investigators from nearly 30 US and foreign law enforcement agencies nabbed 27 others from seven countries suspected of being involved in the crime ring.

Criminals allegedly used three websites to traffic counterfeit credit cards and false identification information, and documents such as credit cards, driver's licences, domestic and foreign passports and birth certificates.

"We believe that the suspects have trafficked at least 1.7 million stolen credit card numbers, leading to losses by financial institutions running into the millions", said acting detective chief superintendent Mick Deats, head of the NHTCU.

1. What is meant by identity theft?

2. Research other cases of Internet-based crime.

▶ Why is computer crime on the increase?

The rapid spread of personal computers, wide area networks and distributed processing, has made information held on computer more vulnerable. The arrival of automated teller machines (ATMs), the Internet and mobile phones has created new opportunities for illegal activity.

Every one of the top 100 companies in the FTSE Index has been targeted or actually burgled by computer criminals. The British police have evidence of 70,000 cases where systems have been penetrated and information extracted. One enquiry revealed three hackers had been involved in making 15,000 extractions from systems.

Banking security experts in the USA have estimated that an average bank robbery nets $1900 and the perpetrator is prosecuted 82 per cent of the time. With computer fraud, the proceeds are nearer to $250,000 and less than two per cent of the offenders are prosecuted.

case study 2
▶ bank fraud

A woman, who opened a bank account using false information, saying that she expected to receive her divorce settlement shortly, carried out a more elaborate fraud. She later returned to the bank and surreptitiously removed all the paying-in slips (used by customers to pay money into their accounts) and replaced them with paying-in slips that she had had specially printed.

These paying-in slips were exactly the same except they had her account number printed at the bottom in MICR characters — just like the paying-in slips at the back of a chequebook.

When reading paying-in slips, the computer looks for the MICR numbers. If there are none, the operator has to type in the bank account number given. If there are MICR numbers, the information is automatically read and not checked.

Money paid in with the fake paying-in slips was paid directly into the woman's account. Customers did not notice any errors until they checked their bank statements. By this time the woman had withdrawn over $150,000 in cash from her "divorce settlement", disappeared and was never seen again.

1. Was the woman's act an internal or external threat?

2. Was the act a crime or malpractice?

Weak points within an IT system ◀

The weak points within an IT system are associated with hardware, software and people. Particular threats include:

- data being wrongly entered into a system
- access to data stored online
- access to data stored offline, such as on a memory stick or CD-R
- viruses, worms and Trojan horses
- data being transmitted using a network, including a wireless network
- internal staff not following procedures
- people from outside trying to see, steal or alter data.

▶ Data entry

Data can be fraudulently entered into the system with criminal intent. A corrupt data entry clerk could purposely enter the wrong account number for a transaction so that an unsuspecting account holder is debited. An employee, operating a system that photographed students and produced college identification cards, was caught accepting bribes from students to input false dates of birth to be printed on to the card. The possibility of such fraudulent acts occurring are called internal threats.

▶ Data stored on computer

Users' personal computers are particularly vulnerable, especially if they are attached to a network. If unauthorised users can gain access to the system they could be able to retrieve, take a copy of or alter data. This could be achieved if a computer user were to leave her computer unattended whilst logged on to a system, with no form of protection.

The use of laptop and palmtop computers produces risks whenever sensitive data is being stored. Such devices are more easily stolen as Case Study 3 shows.

---

case study 3
▶ MI5 laptop theft

(From John Leyden, http://vnunet.com, 24 March 2000)

A laptop computer containing sensitive information on Northern Ireland has been stolen from an MI5 intelligence agent, it emerged today, as security experts warned that not all the information on the device was necessarily secure.

The £2,000 computer was snatched from the Security Service worker as he stopped to help a passer-by in the ticket hall at Paddington Underground station in central London.

■ State three things the MI5 agent could have done to reduce the risk of losing a computer with sensitive and valuable data.

▶ Data stored offline

Data that is stored offline, for example, on a CD-R or a memory stick, is also vulnerable to loss or theft. Disk stores should be kept locked when left unattended. There should be formal clerical systems in place so that details are recorded whenever disks leave the store. The filing and recording system should be maintained rigorously to ensure that files stored on removable media are not mislaid.

▶ Viruses, worms and Trojan horses

A virus is a program that is written with the sole purpose of infecting computer systems. Most viruses cause damage to files that are stored on the computer's hard disk. A virus on a hard disk of an infected computer can reproduce itself onto a memory stick. When the memory stick is used on a second computer, the virus copies itself on to this computer's hard disk.

This spreading of the virus is hidden and automatic and the user is usually unaware of the existence of the virus until something goes wrong. Thousands of viruses exist, with their damage varying from the trivial to the disastrous. Some viruses have little effect. Others delete all your data. Many viruses are distributed by email. You may get an email that says something like this:

Hi! I am looking for new friends.

My name is Jane, I am from Miami, FL.

See my homepage with my weblog and latest webcam photos!

See you!

If you open the attached file or click on the hyperlink, you will probably contract the virus. This sort of virus is very prevalent as Case Study 4 reveals.

---

case study 4
▶ Christmas card virus hits one in 10 emails

(From Robert Jaques, http://vnunet.com, 16 December 2004)

The Zafi-D worm (W32/Zafi-D), discovered earlier this week posing as a Christmas greeting, is spreading rapidly around the world.

IT security experts have reported that the virus is currently accounting for around three-quarters of all virus reports, with some estimates suggesting that the infection is present in as many as one in 10 emails.

Zafi-D, which is believed to originate from Hungary, spreads inside bogus Christmas greeting emails. The emails can use a variety of languages including English, French, Spanish and Hungarian.

■ Give three pieces of advice about how to avoid these viruses.

Activity 1

1. Find out about the latest virus threats at **http://us.mcafee.com/virusInfo/default.asp**.

2. Complete the following table:

Name of virus	What it does

A **worm** is a stand-alone executable program that exploits the facilities of the host computer to copy itself and carries out an action such as using up all the computer's memory and processing capability, forcing the system to close down.

Yet another destructive program type is the **Trojan horse**. This passes itself off as an innocent program. A particular Trojan horse program claims to remove a virus from your computer but, if downloaded, will actually introduce a virus.

▶ Spyware

Spyware is a type of computer program that attaches itself to a computer's operating system. It can severely impede the computer's processing power and take up large amounts of memory. Once installed, spyware can track a user's use of the Internet and display unwanted advertisements or unrequested web sites. These activities can make a Web browser so slow that it becomes unusable.

Spyware, unlike a virus, does not aim to damage the computer. It gets on to a computer as a result of a user action such as downloading a software package or clicking on a button in a pop-up window.

Activity 2

"Masquerading as anti-spyware", "piggybacked software installation" and "browser hi-jack" are all categories of spyware. Find out what each entails.

▶ Networks

Data being transmitted over a network is particularly vulnerable to external threat. The risk of unauthorised access increases when data is transferred over a WAN using public communication links. A line can be tapped to allow eavesdropping of the signal being transmitted along the link. This has been recognised as a real problem for Internet users. Many home users of a wireless network do not lock the network so that anyone with a computer within a working distance can access the network.

Many people will not purchase goods via the Internet because they are concerned about the risk of their credit card details ending up in the wrong hands.

The rapid growth in the use of the Internet, for advertising, for selling goods and for communication, has made many information systems vulnerable to attack.

▶ Internal IT personnel

Security procedures are only as good as the people using and enforcing them. A high percentage of breaches of a company's security are made by its own employees. Sometimes this is due to laziness and not following procedures. Sometimes it is due to dishonesty.

Disgruntled, dishonest and greedy employees can pose a big threat to an organisation as they have easy access to the information system. They may be seeking personal gain and it is not unknown for employees to be bribed to provide information to a rival organisation. Data may be altered or erased to sabotage the efforts of a company. Information about a business may be of great value to a competitor. Industrial espionage does exist in the cut-throat competitive world of big business.

► Hacking

Hacking is a general term used to mean attempting to gain unauthorised access to a computer system – that is, use by a person who has not been given permission to do so. Someone who gains unauthorised access is often called a hacker. Hacking is often achieved via telecommunications links. Many hackers have no specific fraudulent intent but enjoy the challenge of breaking into an ICT system.

Although hackers may have considerable technical expertise, it is possible to find guides on hacking on the Internet. These allow less expert users to become hackers. In some instances, the hacker's purpose in accessing the system could be to commit fraud, to steal commercially valuable data or even to cause damage to data that will have serious consequences for the company whose system it is.

<Exam note>

Try to avoid using the term "hacking" in your answers to questions. The examiner is usually looking for a more specific answer.

Methods of protection ◄

The value of data to an organisation far exceeds the value of the physical computer system it is stored on. It is vitally important for an organisation to keep its ICT systems and data safe from all potential threats. The security measures used by an organisation will reflect the value of the data stored and the consequences of data loss, alteration or theft. Loss of this data could lead to the collapse of the business. Financial institutions, such as banks, need to have the very highest levels of security to prevent fraud.

Precautions must be put in place to prevent security breaches. The types of precautions can be categorised as:

- hardware measures
- software measures
- procedural measures.

▶ Hardware measures

The most obvious way to protect access to data is to lock the door to any computer installation. The lock can be operated by a conventional key, a 'swipe' card or a code number typed into a keypad. Of course, it is essential that any such code must be kept secret. Staff should not lend keys or swipe cards to anybody else. Locks activated by voice recognition or fingerprint comparison offer alternative, stronger methods of security. IBM's ThinkPad is equipped with a fingerprint reader located below the arrow keys, that verifies the identity of a user when he swipes a finger across a tiny sensor.

Additional physical security measures include computer keyboard locks, closed circuit television cameras, security staff and alarm systems. Passive infra-red alarm systems to detect body heat and movement are commonly used, as they are reliable and inexpensive.

▶ Software measures

Identification of users

To make sure that unauthorised users do not access a networked system, all authorised users must be able to be recognised. The usual way to do this involves every user being allocated a unique user identification code and secret password. Only when the identification code and password are keyed in, is the user able to access any of the software or data files.

Network software can be set to:

- only allow passwords that are between 6 and 12 characters
- not allow dictionary words or names to be used as passwords
- only allow passwords that are combinations of letters and numbers
- force each user to change his password after a set time has elapsed
- deny access to a workstation where an incorrect password has been entered three times in a row.

A **network access log** can be kept. This keeps a record of the usernames of all users of the network, which workstation they used, the time they logged on, the time they logged off, which programs they used and which files they created or accessed.

Levels of permitted access

Not all users need to be able to access all the files on a networked system. Access rights should be set up that only allow certain users to have access to specific files or applications. Not all users need to access data in the same way as there are a number of different levels of access that can be permitted:

- Read only: the user can view the data in a file but not alter or delete it.
- Read/write access: the user can modify data as well as view it.
- Append access: the user can add new records but not edit or delete records.
- Delete access: the user has the authority to remove a file or record.
- No access: a user cannot access the file in any way.

Student users of a school or college network have full rights to their own file space. They can read, create, amend and delete files in their area, possibly subject to a maximum file space limitation. No other user, except the network manager, can access their area.

There may be shared areas to which teachers have full access rights but students have read-only access. Teachers can save (write) documents here for student use.

There may be other shared areas used by teachers to store data where the students have no access rights. These may be used for storing important documents such as schemes of work.

Some areas of the disk will be write-protected for everyone except the network manager. For example, only the network manager should be allowed to install new and delete old software or add and delete users. To access very sensitive information, users need to know several passwords.

case study 5
▶ newsagents

Mr Blyth runs a shop in a chain of newsagents, FastNews. The in-shop system has two linked computers, one on the counter and one in the back office. The shop assistants use the counter computer to inform customers of the amount that is owed and enter details of any payments made.

Any details of changes in orders, cancellations due to holidays or closure of accounts are recorded in a black book. New customers fill in a form with details of their address and newspaper requirements.

Mr Blyth uses the computer in the back office to produce round lists for the paper girls and boys. Every night, he uses the data from the black book to update the computer records as well as create records for new customers. Once a month he runs a program that produces sales statistics that he sends to the head office of FastNews.

If any problems occur with the system, FastNews send their technician to the shop to sort it out.

1. Describe the user rights that are needed for each of the three categories of user: Mr Blyth, the shop assistants and the FastNews technician. You may need to consider different categories of data.

2. Discuss the potential internal and external threats to FastNews' ICT system.

► Virus protection: prevention, detection and repair

Although sensible procedures can reduce the likelihood of an ICT system being infected by a virus, software measures are also necessary to ensure protection. Anti-virus software, such as Sophos or Kaspersky, can detect a virus on a computer system and destroy it before it can corrupt data. Free virus protection software AVG is widely installed by home users. The company distributes it free so as to get early warning of new viruses. This sort of software can also prevent viruses getting on to the computer by automatically scanning new files, such as email attachments.

► Spyware protection

A number of anti-spyware software packages are available, such as Microsoft's Defender. The software can provide continual protection against the installation of spyware software and works in the same way as anti-virus protection, by scanning all data coming into a computer from the Internet for spyware software and blocking any that it detects.

Alternatively anti-spyware software can be used to detect and remove any spyware software that has already been installed onto a computer. Scans can be scheduled to be carried out on a regular basis.

► Encryption

Data encryption means scrambling or secretly coding data so that someone who intercepts it, for example, by tapping a telephone line, cannot understand or change the message.

Encryption methods are regularly used to protect important and confidential information when it is stored or being transmitted from one device to another. An example of the use of encryption occurs in the banks' Electronic Funds Transfer (EFT) system. Banks and other financial institutions transfer very large amounts of money electronically. These transfers are protected by the use of data encryption techniques.

The simplest of all the methods of encrypting data uses a translation table. Each character is replaced by another character from a table. However this method is relatively easy for code breakers to decipher. More sophisticated methods use two or more tables. An example of this method might use translation table A on all of the even bytes and translation table B on all of the odd bytes. The use of more than one translation table makes code-breaking relatively difficult.

Even more sophisticated methods exist based on patterns, random numbers and the use of a key to send data in a different order. Combinations of more than one encryption method make it even more difficult for code-breakers to determine how to decipher encrypted data.

▶ Combined hardware and software measures

Some protection methods are a combination of hardware and software

Firewalls

A firewall is used to prevent unauthorised access to a computer system. The growth of always-on connections means that a firewall is now as necessary at home as in a business.

A firewall can operate in one of two ways. A hardware firewall is a dedicated computer that operates between a local area network file server and an external network, such as the Internet. The firewall machine has built-in security precautions to prevent someone from outside accessing data without permission.

A firewall, such as McAfee Personal Firewall Plus, can be installed on a home computer connected to the Internet. The firewall software intercepts any attempt to access locally stored data from outside.

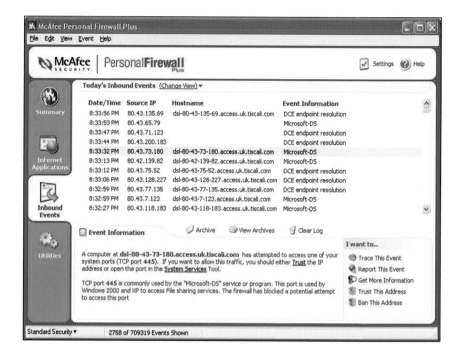

Figure 20.1 McAfee Personal Firewall

The software can be set up to automatically allow access from authorised addresses or sites and to block access from other users.

Biometrics

Biometrics is the name given to techniques that convert a unique human characteristic such as a fingerprint or an iris scan into a digital form that can be stored in a computer or on a smart card.

Biometric data could also be stored about the shape of someone's face, the shape of their hand, their palm prints, their thumb prints or their voice. Biometric data can be used in physical security such as opening doors, accessing computer systems and even accessing cash machines. As this data is unique to the user, fraud and forgery are impossible, in theory.

Biometric data identification can also be stored in documents such as passports and identity cards, so that the holder can prove who they are. IBM has developed an experimental system, FastGate, which uses a hand, voice or fingerprint to ease business passengers through passport control. This system identifies passengers by comparing their fingerprints, voice patterns or palm prints with a digitised record stored on a central database.

▶ Procedures

Around the workplace

Computer systems with terminals at remote sites are a weak link in any system as access to them could provide an intruder with access to the whole system. It is essential therefore that such terminals are fully protected. Computer workstations should be logged off whenever they are not in use, especially if the user is away from his desk. Disk and tape libraries also need to be protected, otherwise it would be possible for a thief to take file media to another computer with compatible hardware and software.

Staff and authorised visitors should wear identity cards, which cannot be copied and should contain a photograph. These are effective and cheap. These security methods are only effective if the supporting administrative procedures are properly adhered to, for example, doors must not be left unlocked and security staff should check identity cards and challenge anyone without one.

Employees working in sensitive areas must be totally reliable. They will often need to be vetted before appointment. Strict codes of practice exist for employees and anyone found to be in breach of these regulations is likely to be dismissed.

The use of an **audit trail** can enable irregular activities to be detected. An audit trail is an automatic record made of any transactions carried out by a computer system. Whenever a file is updated, or a record deleted, an entry will appear in the trail. This means that a record is kept of any fraudulent or malicious transactions or deletions that are made.

If possible, the different stages involved in carrying out a transaction should be divided up so that no one person is responsible for the whole process. This method is also used when a program is being written: no one programmer is allowed to write the whole program, so that producing code that carries out fraudulent transactions will be very hard to achieve without the involvement of other programmers.

Password procedures

Passwords need to be kept private otherwise they have no value. They should be carefully guarded and never revealed to others. A user should take care over the choice of passwords. A password should not be easy to guess. For example, names should be avoided, as should words such as SECRET, SESAME and KEEPOUT. A password should not be too short otherwise it can easily be decoded. Ideally, it should not be a real word but simply a collection of characters, perhaps a mixture of numbers and letters. Some passwords are case sensitive; if so it is a good idea to mix up upper and lower case letters. For increased security, passwords should be changed regularly.

It is essential that a password is never written down. Far too often users write their passwords down in their diary or on a piece of paper which is kept in an easily accessible desk drawer. Even worse is to write the password on a sticky label that is stuck to the screen of the computer.

Virus protection procedures

The risk of getting a virus can be reduced by sensible procedures such as not opening email attachments or using floppy disks if they are from an unknown source. However, it is not enough to purchase and install antivirus software. New viruses are being discovered regularly and it is vital that antivirus software is kept up to date. You will need to register with the antivirus software company to receive updates which are published regularly. Normally you can receive these updates automatically via the Internet.

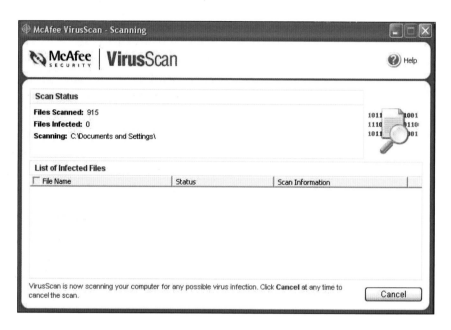

Figure 20.2 McAfee VirusScan

Standard clerical procedures

Loss of data integrity often occurs not as a result of computer malfunction or illegal access, but as a result of user mistakes. For example, yesterday's transaction file could be used to update a master file instead of the correct one for today. This could result in yesterday's transactions being processed twice and today's not at all.

To ensure that such human errors do not occur, very careful operational procedures should be laid out and enforced. Files should be properly labelled and stored in a systematic, predetermined and clear manner. Detailed manual records of the location of files should be maintained.

Write-protect mechanisms

Data can mistakenly be overwritten if the wrong disk or tape is used. Floppy disks and memory sticks are both designed with special slider mechanisms. When the slider is moved to a particular position writing to the disk or the memory stick is prevented. Similarly, certain tapes require a plastic ring to be inserted before the tape can be written to. Care should be taken to write-protect any disk or tape containing data that needs to be preserved.

Activity 3 – passwords

Prepare a set of password guidelines for students at your school or college. The guidelines should address choosing the password, maintaining it and keeping it secret.

Computer Misuse Act 1990

The Computer Misuse Act 1990 was introduced as a result of concerns about people misusing the data and programs held on computers. It allows "unauthorised access" to be prosecuted and aims to discourage the misuse or modification of data or programs.

The Act aims to protect computer users against malicious vandalism and information theft. Hacking and knowingly spreading viruses were made crimes under the Act.

The Act has three sections:

- unauthorised access to computer material
- unauthorised access with intent to commit or facilitate commission of further offences
- unauthorised modification of computer material.

The penalty for each of the three categories is increasingly severe.

▶ Section 1: Unauthorised access to computer material

In this category, a person commits an offence if he tries to access any program or data held in any computer without permission and he knows at the time that this is the case. The maximum penalty is six months in prison and a £5000 fine.

This category applies to people who are "just messing around", "exploring the system", "getting into the system just for the sake of it" and have no intention of doing anything to the programs or data once they have gained access.

It covers the act of guessing passwords to gain access to a system and have a look at the data that is stored. An authorised user of a system may be in breach of this category of the act if she accesses files in the system that have a higher level of access rights than she has been allocated. A student gaining access to a fellow student's area or breaking into the college administrative system is breaking this category of the act. It is an offence even if no files are deleted or changed.

► Section 2: Intent to commit or facilitate further offences

This category covers offenders who carry out unauthorised access with a more serious criminal intent. Persistent offences under section 1 are also included in this second category. Prosecution under this category can lead to a maximum of five years in prison.

Access may be made with an intention to carry out fraud. For example, breaking into a personnel or medical system with the intention of finding out details about a person that could be used for blackmail falls into this category. Another example of an offence under section 2 would be breaking into a company's system with the intention of finding out secret financial information that could be used when carrying out stock market transactions. The information obtained in this way could be used by a rival company.

A further example of an offence in this category would be guessing or stealing a password, using it to access another person's online bank account and transferring their money to another account.

► Section 3: Unauthorised modification of computer material

This third category concerns the alteration of data or programs within a computer system rather than simply viewing or using the data or program. This category includes the deliberate distribution of computer viruses. Prosecution under this category can lead to up to five years in prison.

An offence could include deleting files or changing the desktop set-up. However, the deleting has to be done deliberately and not just by mistake.

The program code could be changed. This could stop a program from running or cause it to act in an unexpected manner. This category also includes using a computer to damage other computers linked through a network even though the computer used to do the damage is not modified in any way.

Alternatively, data could be changed: the balance of a bank account could be altered, details of driving offences could be deleted or an examination mark could be altered.

► Why so few prosecutions?

Even though the Computer Misuse Act has now been in force for a considerable time, there have been relatively few prosecutions under the Act. Organisations are unwilling to make prosecutions, as they fear that such an action will result in a fall in the company reputation. A lack of confidence from their customers could result from any suggestion that the systems and data could be unsafe and vulnerable to misuse.

case study 6
► Computer Misuse Act cases

A temporary employee at British Telecom gained access to a computer database containing the telephone numbers and addresses of top secret government installations. The employee, who had worked at BT for two months, found passwords written down and left lying around offices and used them to call up information on a screen. The employee was guilty under section 1 of the Act as he accessed the data but made no use of it, nor did he tamper with it in any way.

Christopher Pile, who called himself the Black Baron, was the first person convicted under the Computer Misuse Act. Pile created two viruses named Pathogen and Queeg after characters in the BBC sci-fi comedy Red Dwarf. The viruses wiped data from a computer's hard drive and left a Red Dwarf joke on the screen which read "Smoke me a kipper, I'll be back for breakfast ... unfortunately some of your data won't". The Black Baron was guilty under section 3 of the Act as data was altered on a computer's hard disk.

Two 18 year-olds arrested in Wales were alleged to be computer hackers involved in a million-dollar global Internet fraud that involved hacking into businesses around the world and stealing credit card details. The youths were arrested under the Computer Misuse Act 1990; a home PC computer was used for the crime. Apparently they had accessed the credit card databases of nine e-commerce companies and had published the details of thousands of credit card accounts on the Internet. They were prosecuted under section 2 of the Act as they used the data they obtained to carry out fraud.

1. For each of the cases described below, explain which category (or categories) of the Computer Misuse Act has been broken.

   a. A student at a college plays around with the desktop settings of a computer in the IT Centre.

   b. An employee, having used a computer to order some books over the Internet, leaves their credit card details, written on a piece of paper, next to the computer. Someone else finds the paper and uses the details to order some books for themselves, changing the delivery address.

   c. In January 2003, Simon Vallor was jailed for two years having been convicted of writing and distributing three computer viruses. He apparently infected 27,000 PCs in 42 countries.

   d. An 18-year-old man hacked into, and made changes to, a major newspaper's database which cost the newspaper £25,000.

   e. In 2004, John Thornley pleaded guilty to four offences contrary to the Computer Misuse Act having mounted an attack on a rival site and introduced a Trojan-type virus to bring it down on several occasions.

Activity 4 – Computer Misuse Act

Research prosecutions under the Computer Misuse Act. The web site
http://www.computerevidence.co.uk/Cases/CMA.htm provides some useful cases.
Summarise your findings in the format below:

Description of crime	Which section of Act?	Details of sentence

SUMMARY

Computer crime is any crime that has been committed using a computer as the principal tool.

Malpractice is behaviour that is legal but goes against a professional code of practice.

Weak points in the security of an IT system include:

▶ **data being wrongly entered into a system**

▶ **access to data stored online**

▶ **access to data stored offline, such as on a floppy disk**

▶ **viruses, worms, Trojan horses and spyware**

▶ **data being transmitted using a network**

▶ **internal staff not following procedures**

▶ **people from outside trying to see, steal or alter data.**

Weak points can be reduced by:

▶ **physical security**

▶ **internal procedures and codes of conduct**

▶ **encrypting vital information**

▶ **user IDs and passwords**

▶ **access levels**

▶ **firewalls**

▶ **anti-virus software**

▶ **anti-spyware software**

▶ **biometric security**

▶ **vetting potential employees.**

The Computer Misuse Act of 1990 makes the following illegal:

▶ **accessing computer material without permission**

▶ **unauthorised access to a computer to commit another crime**

▶ **editing computer data without permission e.g. spreading a virus.**

Questions

1. Explain, using examples, the differences between malpractice and crime as applied to Information Systems. (4)

2. In 1990, an act was introduced to allow the prosecution of people who accessed computer systems without authorisation.
  a) Name the act. (1)
  b) State, with examples, each of the three sections of the act. (6)
  c) Few companies ever prosecute people under this act. Explain why this is so. (2)

3. John O'Neill takes his company laptop computer home at weekends so that he can do some work at home and so that his son may use the computer for his homework. Describe three threats to the data stored on John's laptop caused by him taking the computer home at weekends. (6)

4. A company offering security services for ICT systems includes the following quotation in its advertisements "You are protected against hackers, viruses and worms, but what about the staff in the sales department?"
  a) Explain how the company could provide protection against viruses and worms. (2)
  b) Describe three ways in which a company's own staff can pose a threat to its ICT systems. (6)

5. Information systems need to be protected from both internal and external threats.
  a) Explain the differences between an internal and an external threat to an ICT system. (4)
  b) Discuss the methods that a company could take to combat the threats. (8)

6. Explain, with reasons, two levels of access that could be given to different categories of users of an online flight booking system. (4)

7. State three methods of preventing unauthorised access to data. (3)

8. All students at a university have a password to access the university's computer network. State three rules that the students should follow to ensure the effective use of the password system. (3)

9. Discuss the relative advantages of hardware, software and procedural measures in preventing security breaches in ICT systems. In this question you will be marked on your ability to use good English, to organise information clearly and to use specialist vocabulary where appropriate. (14)

AQA Specimen paper 2

10. A file containing sensitive data is stored on a computer system. Access to this file is managed by the use of passwords entered at a keyboard and by setting levels of permitted access.
  a) Explain what is meant by
    i) password (2)
    ii) levels of permitted access. (2)
  b) Give two other possible methods of managing access to the contents of this file. (2)

AQA January 06 ICT2

21 Backup and recovery
AQA Unit 2 Section 6

◀

Backup refers to making copies of data that may be used to restore the original if a loss of data occurs. The purpose of backing up is to ensure that if anything happens to the original file, the backup copy can be used to restore the file without loss of data and within a reasonable timescale. Restoring files to the original state, before failure occurred, is called **recovery**.

An organisation needs to put procedures in place that allow lost or corrupted files to be restored by making use of backup copies. These procedures need to be carefully planned and personnel made aware of them; the methods of recovery should be practised so that, when they are needed, they will run smoothly.

Backup is used to avoid permanent data loss and ensure the integrity of the data. It needs to be undertaken on a regular and systematic basis.

case study 1
▶ college students' backup procedures

Masufa, Gary, Mo and Liam are students studying A-level ICT at college. They have to carry out a major piece of course work. As a deadline draws near, they each spend considerable time working on their assignments at home.

Liam keeps a file saved on his hard disk that he calls "project". Whenever he amends the file he overwrites the old version. He intends to copy the file on to his memory stick the next morning and take it in to college.

Mo creates a new version of the file every time he starts a new work session. He calls the files "project1", "project2", etc. He too intends to copy the file on to his memory stick the next morning and take it in to college.

Gary copies the file that he calls "project" to his memory stick at the end of every working session, replacing the existing version.

Masufa keeps a single file called "project" which she saves every few minutes. At the end of a working session, she sends an e-mail to her college mail account with the project file as an attachment.

1. Discuss what would happen to each of the students' work if:

   a. the project file on the hard disk became corrupted

   b. the hard disk crashed

   c. the memory stick was lost.

2. What else could go wrong?

3. Devise secure backup procedures that the students should follow.

case study 2
▶ **Matt loses his photographs**

Figure 21.1 Matt

Matt went for a holiday in Australia. He had some work to complete while he was on holiday so he took his laptop computer with him. Matt is a keen photographer. He had been storing all his digital photos on the hard disk of his laptop for the last 18 months.

While travelling through north-western Australia, his laptop and his camera were stolen. Matt had taken out travel insurance, so his camera and laptop were replaced, but he was devastated to lose all his photographs.

1. What should Matt have done prior to his holiday to ensure that he did not lose all his photographs?

2. What could he have done to make sure that he did not lose his Australian holiday photos?

Backup requirements ◀

Different individuals and organisations have different needs for backup and recovery which depend upon the uses that are made of ICT and the risks to the organisation if data is lost.

The aim of producing file backups is to make sure that if data from a computer system is lost or corrupted for any reason, the files can be recovered and the computer system restored to its original state. The loss might occur due to:

- a hardware fault, such as a hard disk crash
- an incident where a file might, for example, be accidentally deleted or data is lost as the result of a natural disaster, such as a flood or an earthquake
- deliberate actions such as sabotage or terrorism.

When planning a backup procedure the following aspects need to be considered:

- **what** is to be backed up
- **when** is the best time to back up
- **how** the backup should be carried out
- **what media** should be used and where the backup media will be kept
- **whose responsibility** it is to ensure that backups are taken and that adequate testing of the recovery of backed up data is carried out.

Activity 1

Find out more about backup at <u>http://www.backup4all.com</u>. What different types of backup does it offer?

▶ What is backed up?

While emphasis is placed on the need to back up data files, there are other electronically stored files that are crucial to the running of the system.

Without an operating system, a computer is virtually unusable. A modern operating system is complex and requires many stored files. Many of these are specific to the installation and include appropriate device drivers, fonts and control panel settings. Without suitable backup it would be difficult and very time consuming to restore the environment to its original state if the files were corrupted.

Applications, though installed from CD-ROMs, are usually customised once installed to meet the specific needs of the user. Without backup, all such customisation could be lost.

▶ When to back up

It is important to establish when backup should take place. For the many systems that only run during working hours, backup can take place at night. Software can be used that will do this automatically without human intervention. In this case someone must ensure that the backup tapes are changed every day.

A **backup log** should be kept. In this should be recorded the details of all backups taken with details of time, date, the medium used (identified by a volume number) together with the name of the person who carried out the backup.

Frequency of backup

How often should a backup be performed? Obviously the more frequent the backup the less out of date will be the data when it is restored. However, when a backup is performed, processor time is tied up and files can be unavailable for use.

An appropriate balance needs to be found and factors such as the time to restore files in the case of failure as well as the importance and nature of the data will be taken into account. Sales data for a supermarket, which affects orders and deliveries, will be backed up hourly, if not more frequently. User data, such as user names and passwords need only be backed up every week.

Program files do not change very frequently except during development; backups only need to be created when new versions are installed. However, many software applications need to be configured to meet the requirements of the hardware that is installed. They can also be customised to meet the user's specific needs using macros and templates for example. It is advisable to take backups whenever such modifications are made.

Time of backup

When should a backup be performed? If a backup is performed daily, it is likely that the backup will take place overnight when computer system use is much less. However, with more and more organisations running 24-hour operations, this is not necessarily the case.

<tip>

An individual user, such as a home user working on a PC, should not ignore the need to backup. There are a number of measures that can be taken to make sure that, in the event of data loss, important files can be restored.

Many software packages such as Microsoft Word offer automatic backup procedures. When the software saves a file, the previous version of this file is saved as a backup. The old version is automatically renamed before the new one is saved. The old file is usually given the same name but a BAK extension. For example, a file named LETTER.DOC would be backed up as LETTER.BAK. The latest version is saved as LETTER.DOC.

Another software feature that maintains data security is the auto-save feature, which causes the software to save the work automatically every few minutes. This feature is available in Microsoft Word and Microsoft Excel on the Tools, Options menu. If correctly set and the system crashes, even if the user has forgotten to save their work, a recovery file has been stored.

▶ How can data be backed up?

Full backup

Perhaps the easiest way to back up is where a full image of the system is copied to a storage medium such as tape every single day. A different tape is used each day. This ensures that the system can be restored from only one tape. This type of backup is common for small servers that are not in operational use 24 hours a day. The limitation of taking a full backup is that it can take a considerable time to carry out; if a system is running at the same time it is likely to slow down processing.

Differential backup

A differential backup only backs up the files which are different from, or were updated after, the full backup. This allows the full system to be restored with a maximum of just two tapes. The limitation of using differential backup is that when some time has elapsed since the last full backup, the backup of all changed files will be time consuming.

Incremental backup

An incremental backup backs up only the files that have changed since the last backup to the tape. The advantage of an incremental backup is that it takes less time to backup but if a failure does occur it would take more time to restore files to the current position as several tapes may have been used.

For example a business might use an incremental backup strategy as follows:

- Monday night: Full backup on tape A
- Tuesday night: Incremental backup (Tuesday's changes) on tape B
- Wednesday night: Incremental backup (Wednesday's changes) on tape C
- Thursday night: Incremental backup (Thursday's changes) on tape D
- Friday night: Incremental backup (Friday's changes) on tape E

If the business needed to restore on Saturday, all five tapes would be needed.

Unless a file is backed up after every transaction, which is most unlikely to be feasible, a record will need to be kept of all the transactions that have taken place since the last backup occurred. This is called a **transaction log**. In the case of file failure, the latest backup copy would be used to restore the file. It could then be brought up to date by rerunning all the transactions, stored on the log, that had occurred since the backup was made. Of course, the transaction log itself must be backed up otherwise failure could cause a loss of transactions. One important transaction you wouldn't want lost might be the booking of a flight.

In some systems it is necessary to perform transaction log backups every 15 minutes during the day to keep the data entry loss to less than 15 minutes.

Systems that are live for 24 hours a day may need to look at alternative methods of ensuring that data is adequately backed up.

Online backup

The methods described above are used by many organisations and are appropriate for small and medium-sized systems. However, organisations with very large systems that hold vast data warehouses of information need to look at alternative methods for providing backup. As well as having very high volumes of data, the organisations running such systems can afford neither to lose even a small amount of data nor to lose operational time while data is being restored from a backup. It is likely that the system will be running 24 hours a day, 7 days a week. An alternative way has to be found to protect against the hazards of data loss.

One way is to use **disk mirroring**: storing identical data on two different disks. Whenever data is stored to disk, it is stored on both disks. If the main disk fails, the exact data is available on the second disk. The mirror disk does not have to be located in the same place as the first disk. If it is in a different building then the data is protected from disasters such as fire or terrorist attack.

A **redundant array of inexpensive disks (RAID)** is a fault-tolerant system which uses a set of two or more disk drives instead of one disk to store data. By using two disks to store the same data, a fault in a disk is less likely to affect the system.

Remote backup

An organisation can use a remote backup service, similar to the one described in Case Study 3, by transferring files to a remote site using a wide area network. The limitations of using remote backup are that data is vulnerable as it is being transmitted over the network and will need to be encrypted.

<table>
<tr><td>

case study 3

▶ a remote backup service

</td><td>

A wholesale business that buys clothing from abroad and sells to retailers makes use of a company that offers a daily, automated, off-site backup service. Data from the business's computer is automatically compressed, encrypted and backed up after business hours to one of the remote backup company's two state-of-the-art UK data centres.

When the computer network of the business crashed recently, all the current data files were lost. The remote backup company were able to restore the files in a few hours and no data was lost by the business.

The use of a remote backup service has many advantages for the company. Before they signed up for the service they found that their employees sometimes forgot to perform a backup even though they had all the necessary hardware. In fact, they had an earlier scare when they thought that they needed to restore a file that they had backed up only to find that the backup had not completed successfully – the backup tapes were useless. Fortunately very little data was lost, but this scare prompted the owner of the business to take up the remote backup service.

The business no longer has to worry about changing backup tapes, storing backup tapes in a safe place, damaged or lost tapes, maintaining backup hardware, testing backups and training their staff.

1. Why is the data encrypted?

2. List the benefits to an organisation of using a remote backup service.

3. Find out more about remote backup at http://www.backupdirect.net/.

</td></tr>
</table>

It is important that after files have been backed up, the backup is verified; in other words, it must be checked against the original to ensure that it has been copied exactly. If this is not done, the backup files could prove to be useless and the original file would not be able to be recreated.

▶ Recovery procedures

If the original data files or software files are lost or corrupted, the data can be recovered by using programs that restore the data from the backup files.

It is necessary to restore the files in the correct order by following agreed recovery procedures. The file will first be recreated using the most recent full backup and then each subsequent incremental backup file should be accessed, in time order, to update the file. Any transaction log should then be used to restore the most recent transactions.

If the files are used in the wrong order, the restored file will not be correct. Care must be taken in the careful labelling and organisation of backup tapes and disks to ensure that no mistakes are made.

If the hardware has been damaged, for example in a fire, it will be necessary to use alternative hardware on a different site either owned by the company or rented from a supplier.

It is important that staff are trained how to restore data and have a good knowledge of the appropriate procedures. These recovery procedures should have been practised and tested to ensure that the lost files are fully restored to their state before the loss occurred. It is also necessary to test regularly that the backup has worked. It would be frustrating to go to a backup tape after data has been lost and find that the tape is blank.

If a fire or similar disaster occurs, it will be necessary to ensure that more than just the data is restored. Hardware and software will need to be available very quickly. Some very large organisations will have a complete duplicate site with hardware and software that can carry on processing if failure occurs at the primary site.

▶ Backup media

An organisation or an individual user must decide what medium to use for backup. A number of factors will influence this decision. These include:

- the **seriousness** of the consequences of lost data
- the **volume** of the data
- the **speed** at which the data is changing
- how quickly recovery needs to take place if a situation occurs when data has to be restored from backed up versions
- the **cost** of the various methods.

Large businesses with much essential data are likely to use high-capacity storage media such as magnetic tape. They may also use an autoloader – a magnetic tape drive that can fetch tapes from a library automatically and load them. This makes unattended large-scale backups possible. They are frequently used when backing up data stored over a network. The volume of data that can be backed up without the need for human intervention has increased hugely and runs into thousands of gigabytes. It is limited only by the number of tapes that the library can hold.

Backing up small quantities of data

It is not always necessary to have a device that is dedicated to backing up.

Smaller businesses and home users may use storage devices for backup that were designed for archiving or transferring data. Examples of these are:

- **CD-R** and **CD-RW** (compact disk recordable and re-writable) – Normally can store up to 700 MB on each disk
- **Memory stick** – Easy to use, connecting to any USB port and highly portable; can store more than a CD-R and increasing in capacity
- **DVD-RAM** (writeable digital versatile disk) – The same size but greater capacity than a CD-R, up to 4.7 GB
- **External hard drive** – Available in hundreds of GB. They connect to a USB port and can be used to increase disk capacity or as a backup
- **Mirrored USB drive** – Easy to use, connecting to any USB port and highly portable; can take a complete copy of hard disk data – both disks are updated at the same time
- Backing up to the **Internet** – Many ISPs offer their users the facility to back up their data and store it remotely.

Network backup

Larger networked systems usually store all the data on a central file server. This means that individual users do not have to back up as all the data can be backed up automatically. As there are likely to be a large number of files to back up they need to use a backup medium with a large capacity. Magnetic tape, although not used commonly for storing data files is very suitable for backup.

The backup medium used will depend on its capacity, its speed, its cost and the importance of the data.

▶ Physical security of backup medium

As data can be lost due to disasters such as fire, it is essential that backup files are kept in a separate place from the original files. It is a good idea to take backup media off site. Otherwise a fireproof safe should be used.

In some large cities, there exist firms that run a backup service. They come to client organisations at a fixed time, usually daily, and collect the day's backup tapes which they take away to store safely in their own premises. At the same time they provide the client with the tapes, correctly labelled that need to be used for the next day's backups. If the backups are needed to restore the system, the firm will rush the appropriate tapes to the client organisation.

Whose responsibility? ◀

When a network is used, the process of backing up can be centralised and all servers can be backed up from one place. The responsibility of backing up a networked system should be allocated to a specific employee or pool of employees within the organisation. It would be part of their job to ensure that all backups are taken at the proper time, in the proper way and that backup copies are kept in a safe place off site. It is also their responsibility to ensure that recovery procedures are practised from time to time and that careful checks are made to ensure that whatever backup measures are being taken actually work, so that corrupted or lost data can be restored from backed up data at any time. It is important that all recording and playback equipment is compatible.

In most cases, the user is responsible for the backing up of any data that is stored on the hard drive of their own workstation or laptop.

Backup procedures for batch processing ◀

Batch mode is a common method of computer processing used when there are large numbers of similar transactions. It is used when processing such systems as a payroll or utility billing which are run at regular intervals, perhaps once a day, once a week, once a month or once a quarter. All the data to be input is collected together before being processed in a single operation.

In a typical batch-processing system, all the transactions are collected together and entered offline into a transaction file. The records from the transaction file are merged with the corresponding records of the master file and the updated records are stored in a new master file.

The old master file is called the father file and the new version the son. This method of processing produces automatic backup since, if the son file were to become corrupted, it could be recreated by running the update program once again using the same transaction file with the father file to restore the son file.

When the son is in turn used to create a new version, it becomes the father and the old father becomes the grandfather. The number of generations that are kept should be decided upon. Obviously it is necessary to keep the transaction files as well, as they are needed in the process to restore the up-to-date master file.

case study 4
▶ a college backup strategy

A sixth form college runs a network used by teachers and students as well as all administrative services such as MIS and examinations. The network supports over 20 file servers. Backup is an important part of the work of the IT department.

To carry out its backup the department uses a software utility program called BackupExec developed by Veritas, a company specialising in backup solutions.

As little use is made of the college system outside working hours, the department carries out a full backup every night. The backup is initiated by the software and DLT7000 tapes are used to store the backed up files. The tapes are slotted into a 16-slot autoloader.

A rotation system is used. The last week's and previous weeks' tapes are held off site while those for this week are held in the autoloader. The monthly backup is kept for a year.

The backup process starts at 16:30 in the afternoon and is completed around 06:30 the following morning. Jeremy, senior ICT technician at the college, is in charge of checking that the backup has successfully completed when he arrives at work in the morning. The backup software produces a log file which informs him whether the backup:

- completed without errors

- completed with errors

- aborted.

Most mornings the backup will have completed with no errors and a complete failure causing the backup to abort is a rare occurrence. If the backup completes with some files failing to be backed up, Jeremy has to follow up each error. If the backup does fail completely, he has to decide whether or not to risk running without a backup for a day or to slow the network down while the backup operation is repeated.

If a user of the network deletes or overwrites a file by mistake, Jeremy can restore a previous version for them. The backup software keeps a catalogue of the path of all the files it backs up. It will then read through the tape until it finds the required file.

case study 4 (contd)

Files on the servers are not the only things that need backing up. For example, configuration settings for the firewall are stored as a text file. Over time, the settings receive many modifications and it would take many man-hours to recreate if the file were lost.

1. What is the advantage of using an autoloader?

2. Comment on the effectiveness of the college's backup strategy.

3. Why is it important for one person to have the responsibility for backup?

4. Explore the Veritas web site (www.veritas.com) to find out the functions offered by the BackupExec software.

In all other processing modes except batch, data is overwritten as transactions occur, so backup copies of the file will need to be made. With an online sales system, suitable backup facilities must be in place to ensure that all lost transactions can be restored. Consideration of the volume of data involved in the system and the importance of the speed in which restoring from backed up files is necessary must be taken into account.

Many organisations have excellent strategies in place to back up their networked systems and databases but neglect to consider the backup needs of the growing number of laptop computers that are used by employees. Laptops are particularly at risk as they are more vulnerable to theft and potential damage.

An employee's laptop could contain a range of data and programs, such as software applications, highly configured Windows systems, communication setups to connect to the Internet and the organisation's network, documents and other data, and Internet bookmarks, the loss of which could range from annoying to critical.

Activity 2

An insurance salesman uses his laptop to keep records of sales, clients and payments using special-purpose software and to keep in touch with work colleagues through email. He seeks your advice on a backup strategy and asks you to help. Can you suggest answers to these questions?

1. How often should he back up?
2. When is the best time to do it?
3. Should he back up every file every time it is used?
4. What backup medium should he use?
5. Where should the backup medium be kept?
6. How should he log his backups?

▶ Worked exam question

A company should have a backup procedure in case data is lost due to a security breach.

Describe **five** items that need to be considered when reviewing backup procedures.

(10)

AQA Sample Questions

▶ EXAMINER'S GUIDANCE

This question is asking for the factors listed near the start of the chapter, namely:

- *what is to be backed up*
- *when is the best time to back up and how should it be done*
- *how will the backup be carried out*
- *what media will be used and where the backup media will be kept*
- *whose responsibility it is to ensure that backups are taken and that the recovery procedures are tested.*

As there are ten marks allocated to the question, each point above will need to be expanded or illustrated with an example. Remember, you need to write in full, grammatically correct sentences and use ICT technical terminology when appropriate.

▶ SAMPLE ANSWER

When reviewing backup procedures the company needs to decide what type of backup should be taken (1). They might decide that the most appropriate would be a periodic full backup with incremental backups in between (1).

They also need to consider how often a backup should be taken (1); the full backup might be taken only once a week, as it takes a long time, and the incremental backup daily (1).

Another consideration would be the backup medium that should be used (1). A large customer database could be stored onto magnetic tape whereas a small business's emails may be stored onto a USB pen drive (1).

It is important to decide where the backup should be stored (1). In a small business, the owner could store the backup at home while a large organisation may store the backup on another site (1).

The company needs to establish who should be responsible for backup procedures within the organisation (1). An employee or pool of employees should be given the job of ensuring that backups are taken in the agreed way (1).

case study 5
▶ **Porsche cars move to online backup**

Porsche stores vehicle and warranty information as well as customer information. The company has recently moved from traditional tape storage to online backup. Tapes used to be stored off site, so that, in the event of a problem, most historic information was safe. More recent data that had not yet been transferred was vulnerable. If a fire, or similar disaster were to happen on a Friday, then the tapes for Monday, Tuesday, Wednesday and Thursday would still be on site.

The new storage technology will reduce manual error as previously staff have had to swap tapes over.

1. How could the manual swapping of tapes have led to error?

2. What is meant by online backup?

3. What are the advantages of using online backup?

4. What are the drawbacks of using online backup?

The need for continuity of service ◀

ICT systems used by commercial and service industries need to provide continuous service. Inability to access services can cause customer dissatisfaction resulting in loss of custom. Many systems are running continuously, 24 hours a day 7 days a week. Any break in continuity of service can have very serious results.

A few years ago, a failure paralysed a bank's entire national cash machine system, leaving thousands of Christmas shoppers without cash. For more than five hours, customers were unable to obtain money from any automated-teller machine. Bank staff in shopping centres reported big queues as shoppers struggled to get money from cashiers inside branches. The timing could not have been worse: nine days before Christmas, demand for cash is at its yearly peak.

The need for continuity of service is becoming of critical importance to businesses of all sizes as the need for 24-hour service grows. There are two aspects to consider: the need to ensure continuous data protection against disk or other failure, often done using RAID technology; the need to have a replicated system on another site, using disk mirroring, in case of a disaster.

Businesses need to copy vital data in an efficient manner and store it off site so that, in the case of failure, data can be restored quickly so that business can continue. The traditional method of backing up to magnetic tape media which is taken to an off-site location is no longer appropriate for modern systems that run continuously. Restoring data from the tapes is very time consuming and prone to error.

case study 6
▶ **bank recovers data lost on 9/11**

Commerzbank is the world's 16th largest bank, handling US$30 billion in transactions every day. It has offices only 100 metres from the site of the World Trade Center towers, which were attacked by terrorists on 11 September 2001.

Although its building remained standing, hundreds of windows were shattered, equipment was destroyed by smoke and dust, and the offices were evacuated.

Yet Commerzbank was able to resume business in hours. This was because the bank had a disaster recovery site 30 miles away in Rye, New York State.

The company used special mirroring hardware; the disk drives attached to the computers in the World Trade Center were replicated by another set in Rye. Whenever data was changed, a write was automatically made to each of the disks – one for the original and one for the backup. Thus an up-to-date backup of the data, such as customer transactions, financial databases and emails, held at the primary site was held at all times at the remote site.

■ Explain why backup procedures involving full and incremental backups stored off site would not have been adequate for the bank.

case study 7

It was pay day at a food manufacturer and retail group. Anticipating her busiest time of the month, the payroll administrator Marie was at work by 7 a.m. preparing wages data to be transferred from the company's computer to BACS, the system for paying wages directly into banks.

Marie noticed smoke billowing from one of the rooms. All staff had to be evacuated from the offices as fire swept through the building. Everything in the building was destroyed.

Marie collected the backup tapes from another location across town where they were stored. Unfortunately, the computer had been destroyed as well. It took several days to find and configure a suitable computer in a neighbouring business. The 1800 staff were disgruntled that their pay was late in being deposited in the bank.

■ Discuss the recovery procedures that the company could have put in place that would have ensured that their employees were paid on time.

SUMMARY

Backing up refers to the process of copying files. Some software packages offer automatic backup facilities such as:

- backup on saving
- auto save.

Operating systems offer mirror and **RAID** facilities.

The following factors need to be considered when establishing a backup regime:

- what is to be backed up
- when is the best time to back up and how it should be done
- how the backup will be carried out
- what media will be used and where the backup media will be kept
- whose responsibility it is to ensure that backups are taken and that the recovery procedures are tested.

For recovery procedures to be successful:

- backup tapes must be kept securely
- files must be restored in the correct order
- backup tapes must be carefully labelled and organised
- alternative hardware must be available if required
- staff must be trained in what they have to do
- recovery procedures must be practised
- backups must be tested regularly.

A full backup occurs when all the data stored on a medium, such as a hard disk, is copied to another medium, such as magnetic tape.

A differential backup is a variation of full backup where one tape is created that contains a full image of the system and subsequent tapes receive copies of the files which are different or were updated after the image backup.

An incremental backup only stores changes that have been made since the last backup.

Questions

1. A company has procedures to back up the data files held on its computer system on a regular basis.
   a) Explain why recovery procedures should also be in place. (3)
   b) Discuss the elements necessary for a successful recovery procedure. (6)

2. A student is working on an ICT project using the computers at her school and her own computer at home. Describe a suitable backup procedure that the student could use. (4)

   June 2004 ICT2

3. A senior business executive says to his ICT team "As long as you take a copy of our database every few days, we'll be OK". Write a report for the executive explaining why this action would be inadequate. Your answer should describe the procedures that should be in place to ensure that business could carry on in the case of data loss or corruption. (14)

   In this question, you will be marked on your ability to use good English, to organise information clearly and to use specialist vocabulary where appropriate.

4. A company selling camping equipment carries out its business over the Internet. It runs a database system on a network of PCs. The main tasks are the processing of customer orders and the logging of payments.
   a) Explain why it is essential that this company has backup procedures in place. (2)
   b) Describe five factors that should be considered when establishing backup procedures, illustrating each factor with an example. (10)

5. It is estimated that 25 per cent of companies do not have systematic backup procedures.
   a) State what is meant by systematic backup procedures. (2)
   b) Explain why it is necessary for companies to have systematic backup procedures. (2)
   c) In setting up these procedures, one item that has to be considered is which medium to use to store backups. State three other items that should be considered in adopting backup procedures. (3)

What ICT can provide
AQA Unit 2 Section 6 (part 1)

Why are computers used so much?

ICT systems process data and turn it into information. They have become an important part of life today. The main reasons are that they can provide:

- fast, repetitive processing
- vast storage capability
- the facility to search and combine data in many other ways that would otherwise be impossible
- improved presentation of information
- improved accessibility to information and services
- improved security of data and processes.

▶ Fast, repetitive processing

ICT systems process data very quickly – certainly much, much faster than a human could do it. Not only that but they keep getting faster, with processor speed doubling about every 18 months. As computer systems get faster, there are more and more applications for which they can be used.

Accurate, detailed weather forecasts could not be produced without very fast computer systems. They run simulation programs that manipulate data on many different weather factors. To be of any use, a weather forecasting program has to run faster than real time. It would be impossible for human employees to carry out all the calculations required for the forecast in the time required. It is no good getting a forecast for Monday morning's weather on Tuesday, or even on Monday afternoon!

Although there is the possibility of hardware failure or software errors, computer systems are much more accurate in processing data than humans because they can perform repetitive tasks without becoming bored or tired. For example, if a spreadsheet is set up to perform a task such as adding up a list of numbers, it will come up with the same result every time the same figures are entered. A human, on the other hand, is quite likely to get the result wrong sometimes, particularly if they have been working for a long time and are tired.

Many computer applications carry out repetitive tasks. For example, printing bank statements or calculating wages for employees. Paying wages for 500,000 people is no harder than paying wages for five people when using an ICT system; if no such system were in place, many employee hours of work would be needed to carry out the necessary calculations to work out the required information from the data and to produce the payslips. Using an ICT system means that the company requires fewer employees and therefore it saves the company costs.

▶ Vast storage capacity

There are a variety of ways of storing data in an ICT system, all of which store large amounts of data in a very small space. Data can be stored in the computer's internal memory (RAM). This data is lost when the computer is turned off. Data can be stored more permanently using backing storage media such as the hard disk, a CD-ROM disc or a memory stick.

A film can easily be stored on one DVD-ROM disk. Today, even a modestly priced home microcomputer has sufficient storage capacity to enable a user to run a program, such as an action game, that has very realistic animated graphics. Such graphical images require large amounts of storage space. Many thousands of photographic images can be downloaded from the memory card of a camera and stored on a computer's hard disk.

Companies can store detailed records of a wide range of activities relating to the business; this data can be processed to provide a considerable amount of information useful in decision making. This information would not be available if the data could not be stored. A supermarket can store details of every transaction made by a customer, including a record of every item that has been purchased. This data can be processed to produce a variety of information.

Activity 1

The size of computer memory is measured in bytes. A byte can store one character (such as an "e" or a "!"). A kilobyte (kB) is equal to approximately 1000 bytes, a megabyte (MB) is equal to approximately 1000 kilobytes and a gigabyte (GB) is equal to approximately 1000 megabytes. The text of this chapter requires about 200 kB of storage. A typical photograph could require about 5000 kB.

Investigate the storage available on the computer that you use. (You can do this by using the "My Computer" utility in Windows.)

1. What is the capacity of the hard disk? How much is currently being used?
2. Draw up a table showing the storage required for five programs that you use frequently.

Use the Internet to find out the range of storage available for the following types of device and explain what data is stored:

Device	Range of storage	What storage is used for
Mobile phones		
Memory cards for digital cameras		
Memory sticks		
Digital video recorders		

► The facility to search and combine data

ICT systems can search through large volumes of data very quickly. For example, files can be located based on the name of the file, the date of creation or particular text stored within the file.

Consider a computer system, used by the police, that stores details of the fingerprints of hundreds of thousands of people in digitally coded form. When a fingerprint is detected at the scene of a crime a search needs to be made to see if the print matches any that are held. Without a computer system, this task could not be undertaken within a realistic time for all the thousands of stored prints.

Data from different sources can be combined to provide high-quality information in a variety of output formats. In a school or college, a student attendance system collects data on the presence of each student in each of their classes. The data from all classes can be combined and sorted to produce a weekly summary for each student of their attendance in all classes.

► Improved presentation of information

Storing data digitally allows data to be processed into information that can be presented in a variety of ways. The better presentation of items such as invoices, letters and other documents, and a well-presented and accessible website can improve the image of a company and may lead to an increased number of customers and greater profit.

Information should be presented in a format that is appropriate for the intended audience. It is relatively straightforward to produce numerical information in a tabular form or in graphical form using charts and graphs. Using a desktop publishing (DTP) or advanced word-processing package, it is possible to combine graphics and text to produce posters, news-sheets and fliers.

Not all information is printed on to paper. Nowadays many presentations are prepared using software such as Microsoft PowerPoint. Such a presentation could include graphs, tables and text. Presentation software allows for the information to be displayed as a slide show using an LCD projector and a screen.

An LCD projector is connected to a computer and the image appearing on the computer screen is projected. Many lecture rooms or classrooms have a ceiling-mounted LCD projector. This projector is permanently fixed and can easily be connected to a computer, for example, a laptop. LCD projectors can also be connected to video or DVD players where appropriate.

In addition, presentation software gives the option of printing out the presentation so that the audience have a hard copy of the report to take away with them for later reference.

case study 1
▶ **U-Fix-It's information system**

A management information system (MIS) uses data produced by day-to-day operations to create information that can be used in decision-making by management.

A chain of DIY stores, U-Fix-It, maintains a very large database with data on all stock held and customer purchases. For each item of stock data items such as ItemID, Description, SellingPrice and StockLevel are stored.

Every time a customer makes a purchase, a transaction record is added to the database. For each customer transaction, the Date and Time of the transaction and the ItemID and Quantity of each item purchased are stored. For each item purchased, the StockLevel is decreased appropriately. If the customer presents a loyalty card, the CustomerCode is also stored, which allows a link to be made to data relating to the customer that is also held on the database.

The data in the database can be combined in different ways. The company uses the database to produce the following documents:

■ A list of all the stock items that need to be re-ordered

■ A mail shot of letters to all customers who have made purchases of dog beds to inform them of a new product

■ A chart showing the number of sales of selected products during a specified time period

■ A chart showing the number of customers making purchases at different times during the day.

1. What extra information needs to be kept about each product to enable the list of stock items that need to be re-ordered to be produced?

2. What would U-Fix-It need to do to ensure that the mail shot to customers who have purchased dog beds was produced legally?

3. What use would be made of the charts?

4. Describe two further reports that could be created from the database and explain how each could be used by the management.

The use of electronic whiteboards together with LCD projectors has opened up further possibilities for presenters. As well as showing a presentation, an electronic whiteboard allows the user to add notes to the image on the screen and store and print these annotations.

Figure 22.1 Electronic whiteboard used in a classroom

Electronic whiteboards may have a touch screen, so that the presenter can navigate using a finger to move the cursor and double-clicking with taps on the screen. An alternative is to use a special "pen".

▶ Improved accessibility to information and services

ICT systems can link to other ICT systems and other electronic devices, almost anywhere in the world. This facility has increased the number of applications for which computer systems can be used.

Wireless connections are becoming more commonly used. They are more flexible but can only be used over short distances and performance does not match hard-wired systems.

The Internet allows users to communicate with other users worldwide via email and to access and transfer huge amounts of data. A holiday company can store details of all available holidays and bookings that have been made on a central computer system. This is available to travel agents in many different locations who can link their computer system to the central data store. This enables a travel agent both to have access to up-to-date information on holiday availability, and to make bookings immediately, while their client is with them. Alternatively, the booking can be made directly by the customer with no travel agent involved.

The home user has a wealth of information available over the Internet. A site, http://www.ratedtrader.com, provides a householder with access to a range of local traders such as plumbers and electricians. Another site allows users to find the postcode for any address in the UK.

News bulletins and weather forecasts are readily available on a number of websites and can be accessed from some mobile phones and televisions as well as from computers.

Banks offer online services. With online access, an account holder can view both the balance in their account and all transactions that have been made. Standing orders can be set up and payments and transfers made.

Activity 2

ICT allows all the following operations to be performed:

■ Finding out a postcode by accessing the website **www.royalmail.com**
■ Using a tourist information booth with touch-screen display to find out about local attractions
■ Finding a suitable plumber by accessing the website **www.ratedtrader.com**
■ Getting the next day's weather forecast for a town abroad
■ Booking a ticket for a flight from Glasgow to Southampton
■ Setting up a transfer from your current account to your savings account.

For each of the above, describe how the information could have been obtained before the development of ICT systems. For example, it was possible to find out the postcode for an address by visiting a post office where a book was kept with all postcodes in.

case study2
▶ **Ken is off to New York**

Figure 22.2 Ken

Ken is an executive working for an international media company. He lives in South East London and commutes every day to Central London. Today he is flying to New York to attend a conference. He has previously purchased his e-ticket online.

As he eats his breakfast, he checks that there is no delay to his flight by accessing the airline's website using his home computer. He also checks in to his flight using the airline's online facility. He is able to print out his boarding pass, the document that displays his seat number and is required to allow him onto the airplane.

He has booked a taxi to take him to the airport and uses the journey time to check his emails using his PDA. He responds to a few urgent messages.

Ken books in his luggage and then goes to the airline's executive lounge. This has a WiFi hotspot, so he is able to use his laptop computer to access the Internet. He searches for some material relevant to the conference that he is attending.

His flight is called, so Ken logs off his computer and boards the plane.

1. As an alternative to confirming his booking from his home computer, Ken could have used an airport check-in kiosk. Find out more about these kiosks from an airline's website, such as http://www.britishairways.com. List the features offered.

2. List the information that Ken has acquired and the services that he has used via ICT during the morning.

3. Describe the advantages to Ken of the accessibility to information and services that ICT provides.

▶ Improved security of data and processes

The use of ICT can provide greater security. Electronic records are less likely to be lost or damaged than paper records. When data is stored on paper it is not usually feasible to maintain a backup copy. Paper copies are very vulnerable to loss caused by fire, flood or other natural disaster. Electronic copies can be backed up with copies kept in another place so that if original files are lost they can be restored from the backup copies.

When Hurricane Katrina struck and devastated the Gulf Coast region of the US, vital paper documents such as health records and flood maps were lost. When the World Trade Center in New York was destroyed by terrorists, the offices and contents of Morgan Stanley, an international company were lost. However, as the company had a full backup of data and systems located elsewhere, their business was fully restored in a very short space of time. This meant that neither business nor customer confidence were affected.

If paper-based data is to be shared, a document has to be physically distributed. Unless great care is taken, the document could be at risk of being read by an unauthorised person. Data held in an ICT system can be protected from unauthorised access through the use of user identification backed up with passwords.

ICT systems can make it easier to prevent fraud when financial calculations are carried out. Checks can be built in to the software to ensure that the proper process is carried out. For example, when using a POS system in a shop, the amount owing is calculated by the computer rather than by the salesman.

All these characteristics mean that computer systems can be used in many situations today and they provide information of a high quality. Many tasks currently carried out were impossible before the development of computer systems. Accurate weather forecasting, managing international chains of supermarkets and processing financial transactions are all tasks that could not be performed without ICT systems.

case study 3
▶ Public lending library

Figure 22.3 Library computer

A lending library in Wintown has a very large stock of books, DVDs, CDs, audio tapes and video tapes that are loaned to members for a fixed time period. The computer system used to manage the flow of data in the library is linked via a wide area network to the central county computer.

The computer system records details of all loaned and returned items. Whenever a loan is made, the system checks that it does not exceed the member's allowable loan limit; also the member is reminded if they have any overdue items.

An electronic catalogue of all items held by the library is maintained for the use of both staff and borrowers. A borrower is able to check what books are available for loan online from home and to reserve a book for collection later. The central county computer can be accessed to extend the search for a book countywide.

Once a week, a mail-merge program is run that produces letters to be sent to all those members who have items that are over a month overdue.

Every six months, a summary is produced for the senior librarian that provides statistics on borrowing patterns. A list is also produced that contains the details of all items that have not been borrowed during the last period.

1. What is a mail-merge program?

2. Describe the data that needs to be brought together to produce the letters concerning overdue items.

case study 3 (contd)

3. Describe how each of the capabilities in this list relate to the Wintown library lending system:

   a. Fast repetitive processing

   b. Vast storage capacity

   c. Search and combine data

   d. Improved accessibility to information

case study 4
▶ gas billing system

Figure 22.4 Handheld meter-reading collection device

A gas company delivers gas to over half a million customers. Details of all customers are held on a computer system and, every three months, bills are printed out and sent to the appropriate householders. Each bill is worked out from a reading taken from an individual property. Customers are often offered a discount if the bills are sent by email and the account is settled by direct debit every month with an adjustment of an extra payment or rebate at the end of a year.

Details of any payment made are used to update a customer's record. If payment is not made within a defined period, the customer is sent a further bill as a reminder.

Each meter reader has a handheld device that is used to record the number read from the meter. The device already contains details of the customer and house. At the end of the day, the meter reader attaches the device to a connector that is installed at her home. This is then connected via a telephone line to the company's wide area network. Details of the readings made that day are transferred to the central computer system and details of the households to visit the next day are downloaded to the device.

1. What information, apart from the most recent meter reading, is needed to produce a customer's bill?

2. How could bills be produced if there were no ICT system?

3. Describe how each of the capabilities in this list relate to the gas billing system:

   a. Fast repetitive processing

   b. Vast storage capacity

   c. Search and combine data.

case study 5
▶ airline bookings

Figure 22.5 Airline bookings

Every airline maintains a centralised booking system for its flights; travel agents all around the world can access the system to find out about seat availability and make bookings. Travellers can cut out the travel agent by searching for flights over the Internet and booking directly using the airline's website.

Bookings can be made several months in advance. For many large airlines, there will be thousands of flights available for booking at any one time.

Whenever a booking is made, a number of details relating to the passengers is recorded, including dietary requirements. Seat numbers can be allocated, the total cost including airport fees and other charges is calculated and payment is made by credit or debit card.

Many systems allow the travel agent to print the ticket for the customer at the time of booking. Travellers who book over the Internet do not get a ticket but receive a reference to give when they check in. Some airlines allow passengers to print their own boarding cards and check-in documents using a touch-screen computer system.

When the passenger arrives at the airport he checks in, his luggage is booked in and has a label attached that holds details of the destination airport along with an identification code. All the details are recorded on the computer system.

A number of lists need to be produced:

■ a full passenger list for check-in clerks

■ a list showing special dietary requirements for the caterers

■ a list of passengers needing special service together with their seat numbers.

1. What is the baggage code used for?

2. Refer to the list of capabilities given at the start of the chapter. Explain which apply to the airline booking system.

3. What problems could arise from having a booking system that is dependent upon ICT systems?

Limitations of information and communication technology

Although modern-day computer systems have phenomenal capabilities, there are limitations in the use of ICT systems:

- What they can be used for
- The information they produce
- The most appropriate solutions

▶ What they can be used for

The growth in ICT systems has been enormous over the last few decades; 20 years ago many people would have thought that it would be impossible ever to use ICT to carry out many of the tasks that are now carried out routinely using such systems. The development of technology has made the seemingly impossible possible.

However, there are some tasks that are too small to be carried out using ICT and many people feel the need for human interaction in many situations. Although ICT systems can support teachers, doctors and dentists, for example, there will always be a need for some tasks to be carried out by the people themselves.

There has been considerable growth in the development of **artificial intelligence** systems in fields such as:

- **speech recognition**, where software allows a computer to understand a natural human language, such as English;
- **image recognition**, where a computer interprets elements in a visual image, such as recognising a face;
- **robotics**, where a computer is programmed to see and hear so that it can perform like a human;
- **expert systems**, where a computer is programmed to perform like a human expert. Such systems are limited to very small areas of knowledge, for example, an expert system can help a doctor diagnose an illness from a range of symptoms; another can help a mechanic diagnose the fault in a car. Expert systems are expensive to produce and are dependent upon the careful and accurate gathering of the expertise of a human expert.

Artificial intelligence can achieve very little compared with a human.

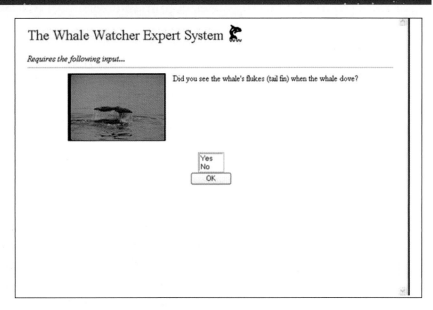

The Whale Watcher Expert System

Requires the following input...

Did you see the whale's flukes (tail fin) when the whale dove?

Yes
No
OK

Figure 22.6 Using an expert system to identify a whale

Access the website http://www.aiinc.ca/demos/whale.html. Run the whale identification expert system.

There are many applications of ICT in education, but although the technology exists it would not be appropriate to completely replace school teachers with ICT systems. Pupils need human interaction to help them learn. Most examination marking needs to be undertaken by a human expert. Multiple-choice questions can be marked using a computer system, but other types of question require the judgement and interpretation of a skilled human.

Although technological advances have been made, financial and other constraints limit what an ICT system can be used for. For example, a home user may not be able to produce high-quality photographic prints because he does not have an appropriate printer. Users in some parts of this country are currently unable to access the Internet using broadband technology.

▶ Information they produce

The quality of the information produced by any ICT system is only as good as the data that has been input. If the data is inaccurate, out of date or insufficient then the information produced will not be reliable. In many systems, the quality of the data input is dependent on human performance.

It is vital to have appropriate data-control mechanisms that detect human errors if an ICT system is to produce reliable and accurate information. These controls should be incorporated into the system when it is being designed. Validation can be used to detect any data that has been entered that is not sensible. However, it is sometimes difficult to prevent or pick up all errors.

When referencing an individual's criminal record, only convictions made in the UK are recorded. This could cause problems for employers who could employ a person who has committed crimes that make him unsuitable for the job.

Some off-the-shelf software in areas such as school administration limits the range of reports that can be produced to a fixed number that have been thought to be useful by the team designing the software. A school manager may be unable to obtain the information in the form that he wants, even though the appropriate data has been collected.

▶ Not always the most appropriate solution

The development of some very large systems has been problematic and led to failure or inappropriate systems. In a number of cases, the implications of the new systems have not been explored fully.

- When a new ICT system was introduced by the Child Support Agency in 2004, only one in eight single parents received the correct amount. Applicants had to wait between 15 and 22 weeks for their first payment. The implementation of the ICT system, which cost £456m, was two years late.
- When new teachers are appointed, a school has to carry out police checks to ensure that they are suitable to work with children. When a new ICT system was introduced, a number of schools were forced to close due to delays caused by the system.
- A computer system designed to administer a training scheme for young people was found to be so vulnerable to fraud that the whole scheme had to be abandoned.
- A newly implemented ICT system to process passport applications failed. Passports were delayed so long that people had to cancel holidays. In the end, the Passport Office had to employ extra staff to deal with the problem caused by the failed system.

Poorly designed systems do not achieve what they set out to do. A system can be poorly designed as a result of inadequate investigation. Perhaps some possible data inputs have not been considered or the results have been calculated in the wrong way. Remember the computer is only a dumb tool and only does what it is programmed to do. If the programming is wrong, the output is wrong. Software errors can make a system fail or behave in an unpredictable manner. This can happen because the software has been rushed onto the market without being fully tested, often for commercial reasons, perhaps to beat a competitor.

Individuals and businesses rely heavily on ICT systems. This can lead to problems if something goes wrong with the system. Companies need to prepare detailed and expensive disaster recovery plans that can be put into action if hardware or data is lost due to unforeseen circumstances such as fire, flood or terrorist attack. Without a plan that will enable an emergency system to be up and running very quickly, a company can lose vital business. This could lead to the company's collapse.

In August 2004, the Post Office's Electronic Benefits Transfer system, which automates payments of the state pension, crashed and was out of action for five hours. The system failure meant that pensioners were unable to collect their pension as they had expected to. The system provides each pensioner with a swipe card and PIN number. This system replaced paper benefits books. Some pensioners who were unable to obtain their money were given an emergency payment hotline number; they were unable to get help as the lines were jammed due to the volume of attempted calls.

Activity 4

Visit a newspaper or magazine site on the Internet such as http://www.computerweekly.co.uk/ or http://www.computing.co.uk and search for "computer system error". Find three examples. For each example, determine the cause of the error.

case study 6
▶ hospital patient records

A new computer system costing £6.2 billion is being installed in the NHS. The system will store records for 50 million patients. By 2010, medical staff will be able to access and update information on their patients from every doctor's surgery and every hospital in England. Wireless access will allow doctors to access the system as they visit different hospital wards.

A number of different medical personnel will use the system:

■ A clerk records details of appointments and visits to out-patients clinics.

■ The ward clerk enters details of a patient's stay in a ward.

■ The radiographer stores details of X-rays.

■ Laboratory staff enter results of blood and other pathology tests carried out.

case study 6 (contd)

- Ward nurses and doctors enter details of medications issued and treatments undertaken.

- GPs can refer to all the information relating to a patient under their care.

When a patient makes an appointment with a doctor at an out-patients clinic, the information is used in several ways. It is used to generate a letter to the patient confirming the appointment and to produce an appointment list for the doctor for the particular clinic session.

Whenever a pathology test is carried out, the results that are stored on the patient record can be viewed by the patient's doctors.

1. State two advantages of using the ICT system described over a manual one.

2. Why will each patient need to be allocated a unique patient number?

3. Describe how each of the capabilities in this list relate to the hospital patients system:

   a. Fast repetitive processing

   b. Vast storage capacity

   c. Search and combine data

   d. Improved accessibility to information

   e. Improved presentation of information

   f. Improved security of data

4. There was considerable controversy over the idea of copying of patients' medical records from local GP records to the national database. Many people felt that this would affect a patient's right to confidentiality.

   a. Research the issue using a newspaper website such as for the Guardian or The Times.

   b. Discuss reasons why keeping a central record would be of benefit to a patient.

   c. Discuss the problems that could arise from medical data being held on a national database.

Discuss any further problems that can arise when such a large-scale ICT project is undertaken.

ICT systems provide:

- ► fast, repetitive processing
- ► vast storage capability
- ► the facility to search and combine data in many ways that would otherwise be impossible
- ► improved presentation of information
- ► improved accessibility of information and services
- ► improved security of data and processes.

However, there are limitations to the use of ICT systems:

- ► in what they can be used for
- ► in the information they produce
- ► in whether they are the most appropriate solutions.

Questions

◄

1. When Mrs Brown received her gas bill she found that it was for £10,000, which she knew was not correct. When she telephoned the gas company to complain, the explanation she received was that "the computer had got it wrong". Describe a more likely explanation. (2)

AQA June 2005 ICT1

2. Most employers now use computer systems to calculate their employees' wages and pay them straight into the employees' bank accounts. Describe four capabilities of ICT which makes electronic payment advantageous to the employer. (8)

3. Over half the world's emails are believed to be spam. This is unrequested advertising sent to collections of email addresses offering to sell you anything from medication to a university degree. Describe three capabilities of ICT which have made spam so prevalent. (6)

4. A hospital is planning to provide all its doctors with handheld computers. These computers will be able to be used anywhere in the hospital to access patient information and order medicines.

   a) List two benefits to the hospital of introducing this system. (2)

   b) Explain one disadvantage to the doctor of introducing this system. (2)

Types of processing
AQA Unit 2 Section 7 (part 2)

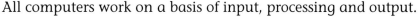

All computers work on a basis of input, processing and output.

Different computer systems require different modes of operation that determine how the computer is used. Sometimes it is appropriate or necessary to process each occurrence of data as it presents itself, whilst in other situations it is more appropriate, efficient and cheaper to collect a large amount of data before processing it all together.

For example, if you go to a bank's Automated Teller Machine (ATM or "hole in the wall") to find out the current balance of your account, you want a response there and then, within a few seconds. You would not be happy to wait for 50 other people to make a similar request before yours was dealt with!

On the other hand, the most efficient way to enter the responses to a survey involving several hundred people is to collect them together and read them automatically, by using an input device such as an **optical mark reader (OMR)** or **optical character reader (OCR)**.

The three modes of operation to be considered are:

- **batch**
- **transaction**
- **interactive**.

There is some overlap between them; most transaction-processing applications are also interactive.

A number of factors will determine the mode of operation for a particular system. These include hardware availability, the volume of data, the required response time and the nature of the system.

Before investigating in more detail, it is necessary to define some of the terms that will be used.

Term	Definition
Master file	A main file of records relating to an ICT system. The data in this type of file does not change very frequently. For example, a banking system would have a master file of account details made up of fields such as name and address, with one record for each account. A stock-control system for a retail business would have a stock master file where each record holds details relating to one item of stock.
Transaction	A single change to a record held in the master file. The record of a deposit or withdrawal of an amount of money into or from a bank account would be a transaction. The details of a change of customer's address would be another.
Transaction file	A collection of transaction records built up over time and used to update a master file of changes.

Batch processing ◀

Batch mode is a method of computer processing used when there are large numbers of similar transactions that can be processed together. All the data to be input is collected before being processed in a single operation. For example, when a survey has been carried out with the responses recorded on forms, the data from all the forms can be input into the system (perhaps using an OCR or OMR device) and, when it has all been entered to create a transaction file, the necessary processing can take place.

Early computer systems, using large mainframe computers, all operated in batch mode as the hardware did not make another mode of operation possible.

There are a number of tasks that are best carried out in batch mode. Batch processing is the most appropriate and efficient mode to use when a large number of records in a master file need to be updated.

In a typical batch processing system:

1. All the transactions are collected together offline then a transaction file is created.
2. This transaction file is sorted into the same order as the records in the master file.
3. Each record from the transaction file is read with the corresponding record of the master file.
4. An updated record is created and stored in a new master file.

These steps are repeated for every record in the transaction file.

The old master file is called the **father** file and the new version the **son**. When the son is in turn used to create a new version it becomes the father and the old father becomes the **grandfather**. A number of generations can be kept for backup purposes. As long as the relevant transaction files are also kept, the current master file can be recreated.

▶ Batch processing: a billing system

An example of a typical batch processing operation is an electricity billing system. The computer already has much of the data required to calculate the bill stored on disk in the master file. The data fields stored will include:

- customer number – the unique code used to identify the customer
- name of customer
- address
- last meter reading – to calculate the number of units used and the charge

- amount of electricity used in the last four quarters – for checking that this quarter's usage is sensible
- special instructions for the meter reader, for example, "the meter is round the back of the house".

The processing is as follows:

1. The meter reader enters the meter reading into a palmtop computer. At the end of the day, he takes the palmtop to the office.
2. The data from the palmtop is used to create a transaction file. Each transaction record needs two items of data: the customer number, so that the correct record in the master file can be identified, and the current meter reading.
3. The records in the transaction file are not in any order; they have to be sorted into the same order as the master file.
4. The two files are read through together, record by record, the customer number being used to match transactions with the master file records.
5. Each time a match is found, the details of the bill can be calculated. A record is created for a new version of the master file, which will contain data updated from the transaction record. This new version of the master file will be used in three months' time when the next reading of the meter is made.
6. When the new version of the master file is created, the old version is still kept as it provides a backup. If the new version were lost or corrupted, it could be recreated by running the update program again using the old master file and the transaction file.

Using batch processing for this billing system means that details of only one customer are processed at one time – the whole file does not have to be loaded in – so that the computer does not need a large amount of memory. Very large volumes of data can be processed efficiently in one run by using batch processing. Each input file is read through just once from start to finish. Once the program is running there is no human intervention.

All the data has to be collected and entered at the start and no more can be entered once the system is running. This means that information can be out of date and new data can only be entered with the next batch.

You cannot perform an immediate search for information as response time for a specific query is slow.

Batch processing is suitable for payroll or billing systems, which are run once a week, once a month or once a quarter. It is not suitable for a system that needs an immediate response, such as an enquiry or booking system.

Outline the stages involved in a batch processing payroll system for a large organisation that inputs the number of hours worked by employees and produces pay slips.

Transaction processing ◀

Transaction processing is a mode of operation where data for each transaction is entered at source and processed immediately.

In a building society, customers can be issued with passbooks. When wishing to make a deposit (put money into their account), a customer can take their passbook, together with a completed deposit transaction form, to a till at any branch office. The building society clerk enters the details of the transaction (account number and amount of deposit) using the keyboard of a networked computer. The account details record in the master file is updated to reflect the transaction; the old values are overwritten. The passbook is placed into a special printer and the details of the transaction and the new balance are printed.

In a booking system, for holidays or cinema tickets for example, the system deals with each booking as it is submitted. Each transaction is completed before the next can begin. The same seat cannot be booked twice.

Unlike in batch processing, the master file is updated with the data from the transaction immediately. Only one record of the master file is accessed, so the file must be organised in a way that allows each record to be located directly and independently. The great advantage of transaction processing is that the master file is always kept up to date. However, master file backup becomes more of a problem as there is no automatic production of backup files as there is when batch processing mode is used.

Many systems use a combination of processing in batch and transaction modes. The building society needs to produce annual account statements for members who may need the details for tax purposes. A printed summary is needed for each account. This is most efficiently produced automatically, in batch mode. It would be a long task if transaction processing were used, as a clerk would need to enter each account number in turn to initiate each transaction.

Interactive processing

Interactive processing involves the user "having a conversation" with the computer. The user may use a number of input devices, such as mouse, graphics tablet, joystick and a keyboard. The response time for interactive processing must be sufficiently fast to avoid frustration for the user. Ideally the user should think that the response is immediate.

Purchasing tickets via a touch screen at a railway station or using an automated teller machine (ATM) to access bank account information are both examples of interactive processing.

Much home computer use involves interactive processing. Most computer games require interactive processing, as do design and drawing packages.

Most transaction-processing systems also involve interactive processing. For example, consider a mail-order company that processes telephone orders on a computerised system as each order arises. This is interactive processing as the operator enters the data directly into the computer and responds to prompts for more data. The order is dealt with whilst the customer is on the telephone; the operator checks availability of each item for the user and confirms the price. The transaction is completed before the operator deals with another customer. Thus the system is based on transaction processing.

Activity 2

For each of the systems below, identify and justify the appropriate mode (or modes) of operation.

- Airline reservation system: A user sitting at a terminal types in details of the customer's request; details of suitable flights are displayed. A booking can then be made and is processed immediately.
- Play-a-Toy plc: Play-a-Toy is a large manufacturing company with around 6000 employees. On the last working day of every month, the computer system automatically transfers wages to the employees' bank accounts and produces individualised pay slips.
- British Telecom: Every three months, a bill is calculated for each of its thousands of customers and printed onto pre-printed paper for posting.
- The home user: Mary has a computer at home with a modem connection and access to the Internet. On occasions, she uses a browser to surf the Internet and download interesting articles.
- Examination marking: When students take multiple-choice examinations they fill in OMR sheets with their choices marked in pencil. The sheets from every school are sent to the examining board, who read the data into a computer system that calculates the results. Individual student marks are summarised on a list for each school.

- HSBC Bank mortgages: When a customer requests a mortgage, an advisor sits down with them in front of a computer. A program guides the advisor through a series of questions for the customer and the answers are entered. At the end, the program produces details of any possible mortgage offer.
- American Express: Each month a bill is produced for each customer. It lists the details of all transactions made in the last four weeks, together with details of any payments made.
- Zap 'Em: Zap 'Em is an arcade game in which buttons and joysticks are used to position a zapper gun to shoot down deadly aliens. These aliens are moving around the screen demolishing innocent lemmings. As soon as a user hits a target, an alien is removed.

case study 1
▶ Howse, Hulme and Byer, Estate Agents

Howse, Hulme and Byer is a chain of estate agents with 50 branches around the country. The current data-processing system, which has been in place for a number of years, is based on batch processing. The main categories of transactions that take place at the branch offices are:

- details of new potential purchasers
- details of new houses for sale
- details of sales.

A branch employee writes details of a transaction onto a pre-printed form. These forms are collected and, three times a week, are sent to the head office where they are put together with forms from all the other branches. A data-entry clerk keys the transaction data into a PC. The data is entered again by a second clerk so that any mistakes made by the first clerk are highlighted.

As data from all the forms is keyed in, it is stored on disk in a transaction file. This file is then input into a validation program that produces a valid transactions file as well as an error report that contains details of mistakes in the transactions. The reports are sent back to the branch offices so that the transactions can be corrected and resubmitted.

The file of valid transactions is then sorted into the same order as the master file and used to update the old master file by creating a new version that includes the changes resulting from the transactions.

An updated list of houses available, ordered by area and price, is sent to each branch. Lists of newly available houses that meet their requirements are sent to prospective buyers.

Howse, Hulme and Byer are considering changing from a batch system to an interactive system. To do this, computers will need to be installed in each branch office. These will be linked to a central computer. Transactions will be processed as they occur, the details being keyed in at the branch offices.

case study 1 (contd)

The new system would bring a number of advantages. As the transactions would be entered into the system as they occur, the information available would be more up to date; details of new houses would be available in the branches as soon as they are received. The details of all houses for sale would be available in all branches as soon as the transaction was complete. The confirmation of a sale would be immediately recorded, thus preventing potential purchasers being shown the details of a house that is no longer available. Any data-entry errors that are made can be corrected immediately and do not have to wait until the next batch is run.

However, there are some issues that need to be considered before the system is updated. Obviously, there will be hardware and software cost implications. Currently, processing takes place when there is little other use being made of the computer system. At present, all data entry is performed by trained staff in the head office. The new system will demand extra skills of branch employees and training will be needed in the use of the new system. There will be a greater security risk as computer records, at present only available at the head office, will be available at all the branches. Backup files, which are produced automatically in a batch-processing system, will require more complex organisation.

- Draw up a table of the benefits and drawbacks of moving to a new system based on transaction processing. Refer to the text and add some ideas of your own.

▶ Worked question

A college uses a computer-based, batch-processing system for keeping the students' records. The students provide their details, or changes to their existing records, on pre-printed forms. The completed forms are collected into batches ready to update the master files. These occur every night at certain times of the year and once a week at other times.

a) Explain what the term "batch processing" means. (3)

b) i) Give one advantage to the college of batch processing. (1)

 ii) Give two disadvantages to the college of batch processing. (2)

c) The college decides to install a transaction processing system in which student records are keyed in online by a clerk. Explain what the term "transaction processing" means. (3)

▶ EXAMINER'S GUIDANCE

Although this question is based around a scenario that you may not have studied, it is easy to adapt what you have learned from a similar scenario, for example, case study 1.

Part (a) of the question requires you to write a definition of batch processing which can be taken from the text on page 284. There are 3 marks available for your answer which means that you have to make 3 distinct points. The following answer actually has 4 points, so if you left one out you would still gain full marks.

▶ **SAMPLE ANSWER**   All the changes to the student records are collected together to form one batch (1) over a day, when the system is busy, and weekly otherwise (1), to be processed in one computer run (1) without any human intervention (1).

▶ **EXAMINER'S GUIDANCE**   *Part (b)(i) is only looking for one advantage to the college. The advantages are similar to those of the estate agency in case study 1.*

Part (b)(ii) can again be worked out by looking at the question and comparing it with case study 1.

▶ **SAMPLE ANSWER**   i) Processing can be done at night when the computer system is quiet (1), as the college is closed; it also requires few staff to run and has few hardware requirements – just one computer (1).

ii) Details may be out of date for up to a week (1) and error corrections may take a further week (1).

▶ **EXAMINER'S GUIDANCE**   *(c) Again this is a formal definition (see page 286).*

▶ **SAMPLE ANSWER**   Transaction processing deals with each set of data as it is submitted (1): each transaction is completed (1) before the next is begun (1).

---

case study 2
▶ booking a seat at the cinema

A cinema sells tickets for numbered seats. The cinema maintains an ICT system to deal with all bookings.

A customer can buy tickets in person either just before the film is to be shown or in advance. When the salesperson selects the appropriate film and the date and time of performance, a layout of the cinema is displayed on a screen showing which seats are still available for booking. The customer can then select the seat or seats required and the system records the booking and produces the tickets.

The customer can also make a booking by telephone or online.

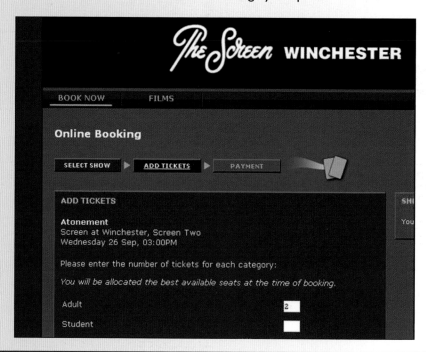

Figure 23.1 Booking screen for cinema tickets

1. What types of processing are involved in the seat-booking system? Justify your answer.

2. A key requirement of a seat-booking system is that seats cannot be double booked. Explain why this is important and describe how the type of processing used will make sure that this requirement is met.

3. Under what circumstances would a customer choose to make a booking:

   a. in person

   b. by telephone

   c. online?

4. What kinds of information could the seat-booking system provide for the managers of the cinema?

case study 3
▶ **buying a train ticket by mobile phone**

A new technology has been developed that delivers train tickets to mobile phones. A barcode is sent to a mobile phone via a text message. The barcode can be scanned at the station ticket barrier.

1. What are the benefits to the passenger of buying train tickets this way?

2. What are the benefits to the rail company of passengers buying train tickets in this way?

3. Search the Internet to see where this technology has been implemented.

4. What types of processing are involved in this system?

SUMMARY

Three modes of operation can be identified and these are shown in the table below:

Processing mode	Description	Example of use
Batch	All the data to be input is collected together before being processed in a single operation.	Invoicing systems; payroll systems
Transaction	Transactions are accepted from outside sources and each transaction is processed before another one is accepted	Online order processing systems; booking systems
Interactive	Data is processed in conversational mode	Designing; games; instant messaging

These modes are not mutually exclusive. The appropriate choice of mode depends upon the nature of the application.

Questions ◀

1.
   a) Explain, using a suitable example, what the term "batch processing" means. (3)
   b) Explain, using a suitable example, what the term "transaction processing" means. (3)
   c) Explain, using a suitable example, what the term "interactive processing" means. (3)

2. Explain, using a suitable example, what the term "transaction" means. (2)

3. A sixth form college enrols its students on to courses over a period of five days. It is considering converting its current daily batch-processing system to a transaction-processing system.
   a) Explain the terms "batch processing" and "transaction processing" in the context of the college enrolment process. (6)
   b) Discuss the benefits of running a transaction-processing system. (8)
   c) Describe any limitations of a transaction-processing system. (4)

24 *The impact of ICT*

AQA Unit 2 Sections 8 and 9

A number of factors affect the way in which ICT systems develop. These can be categorised into the following aspects:

- **cultural** – relating to the behaviour patterns, beliefs and ideas of the population of the country
- **economic** – relating to the financial issues and wealth of a country, social group or organisation
- **environmental** – relating to the impact of changing the ecology of the world
- **ethical** – relating to the accepted principles of right and wrong
- **legal** – relating to the law
- **social** – relating to human society and how it works.

It can be seen that there is overlap between the categories above.

ICT systems also have consequences that affect the lives of individuals and of society as a whole.

Cultural issues

Different groups of people have differing expectations of what ICT can provide. This can depend on where people live, their economic background and their age and experience.

The access or lack of access to ICT in education can affect an individual's readiness and ability to make use of ICT systems.

In the UK, problems with ICT terminology for non-English speakers can act as a barrier to effective use of systems.

Some very elderly people find the forced change in the way they carry out daily tasks confusing and threatening. For example, many shops no longer accept cheques as a means of payment, expecting non-cash payments to be made using "chip-and-PIN" cards. A person, perhaps suffering from mild dementia, who is unable to remember their PIN or does not know how to key it in, could have their card refused at a shop and thus be unable to make their purchases.

An increasing number of government services are now available online. For example, it is possible to register as an elector or access benefits online. Many products are cheaper to buy online. Unfortunately, only 28% of people over the age of 65 have home internet access, compared to a UK average of 57% of households.

Economic issues

The use of the latest information and communication technologies can provide competitive advantage to companies. Well planned and effective systems can reduce costs, carry out operations that could not have been tackled before and increase the reputation of the company.

However, in some situations the financial cost of implementing an ICT solution is too high to make commercial sense in spite of any benefits that might come from the new system. Managers should carry out a **cost–benefit analysis** before starting any new ICT project to ensure that the benefits outweigh the costs.

As more and more access to information and services has become available online, people who cannot afford access have become disadvantaged. An individual needs access not only to a computer but also to fast, reliable broadband – as the use of the Internet becomes more sophisticated, greater volumes of data need to be transferred, for example, for pod casts and video streaming. This disadvantage can affect the life of an individual both in leisure activities and in the ability to work online for employment.

People who do not have the skills to use the available technology are also disadvantaged. This disadvantage can affect the life of an individual. Much vital information for every sphere of life such as education, career and entertainment is increasingly provided via the Internet. The links between poverty and lack of access to ICT have become known as the **digital divide**. The term refers to the gap that exists between those people who have regular access to ICT and those who do not.

The divide can come between:

- the young and the old
- those who live in developed countries and those in the developing world
- those who live in towns and those who live in the country
- the rich and the poor
- those who live in mainstream society and the socially excluded
- those with high and those with low ICT skills.

Access to the Internet could improve the quality of life for many people in the developing world. Currently, the telephone infrastructure is not adequate to support reliable Internet access in many places and, when access is available, the cost is high.

Activity 1

Describe how cheap, reliable access to the Internet in the developing world could help:

■ an individual
■ commerce
■ the health of the population.

It has been proposed that homeless people in the UK should be given broadband Internet access. It is felt that this could provide a lifeline for people on the streets. Access could be achieved by providing homeless hostels and community centres in deprived areas with broadband access.

Environmental issues ◀

The increase in the number of ICT systems is having a marked effect on the environment. It has lead to an increase in resource consumption.

Technology is advancing very quickly. Every year the speed and memory capacity of computers increases substantially. Mobile phones with new features are regularly launched. The consumer and industry are left with the problem of disposing of the old equipment.

There are several problems to consider. The first, one of security, concerns the potential problem of personal or valuable company data becoming accessible to inappropriate people. This issue was discussed more fully in Chapter 20.

The second problem lies in the way in which old equipment is disposed of. An EU directive, the Waste Electrical and Electronic Equipment Directive (WEEE Directive) has been drawn up. A description of the directive on a government website states that "it aims to minimise the impact of electrical and electronic goods on the environment, by increasing re-use and recycling and reducing the amount of WEEE going to landfill. It seeks to achieve this by making producers responsible for financing the collection, treatment, and recovery of waste electrical equipment, and by obliging distributors to allow consumers to return their waste equipment free of charge". See http://www.dti.gov.uk/innovation/sustainability/weee/page30269.html for more information about the directive.

There are many reports that computers are being dumped in the developing world where they are releasing toxic materials into the environment. A recent report describes a group of villages in south-eastern China where computers from the US are to be found lying in large numbers along rivers and fields.

case study 1
▶ computer disposal hazards

An article by Yo Takatsuki, business reporter for the BBC World Service, states that computers and mobile phones contain a large variety of chemicals and plastics which can cause serious harm if not dealt with correctly. It lists the following hazardous waste materials found in ICT devices:

■ **lead** in cathode ray tubes and solder

■ **arsenic** in older cathode ray tubes

■ **antimony** trioxide as a flame retardant

■ **polybrominated** flame retardants in plastic casings, cables and circuit boards

■ **selenium** in circuit boards as a power-supply rectifier

■ **cadmium** in circuit boards and semiconductors

■ **chromium** in steel for corrosion protection

■ **cobalt** in steel for structure and magnetism

■ **mercury** in switches and housing.

1. Investigate the harmful effects of three of the listed substances.

2. Many charities offer a recycling service for mobile phones. Find out about such a scheme.

3. Explain why so much ICT hardware is discarded by users.

4. Find out more about the WEEE directive. What effect is the directive having on the disposal of items in the UK?

Many people are worried that mobile phone masts that are located close to housing or schools can have a harmful effect on health. The growth in the use of ICT technology has led to an increase in electricity consumption; producing electricity involves using up the earth's resources.

Ethical issues ◀

As huge databases are built up, it is possible to link together data on an individual from a variety of sources to produce a detailed profile of the individual. Any information gathered could be used to the disadvantage of the individual involved. It would be technically possible for an employer to have access to the health and education records of their employees.

The UK government is planning to bring in identity cards by 2008 to strengthen national security and protect people from identity fraud. They believe that the cards will tackle illegal working, stop people from claiming free services to which they are not entitled and make it more difficult for criminals to maintain multiple identities.

Each card will contain personal information – the holder's name, address, gender and date of birth together with a photograph. It will also incorporate a chip that will hold biometric data, for example details of a person's fingerprints, which are unique for every individual and therefore difficult to forge.

Opponents of identity cards are concerned about the high cost of implementation and the dangers to civil liberty if the cards were to be made compulsory.

Activity 2

1. Describe the current progress of the implementation of identity cards in the UK.
2. What is now to be stored on the cards?
3. Explain how the use of identity cards could reduce illegal working and fraudulent claims for benefits.
4. Explain why some people fear that the use of cards could erode civil liberties.

There is evidence that the Internet has led to a number of undesirable practices that target vulnerable people in society. Pornographic images are very easy to access via the Internet and can be viewed by anyone, including children unless suitable safeguards are put in place on a home computer. The use of **spamming** has grown; hundreds of thousands of Internet users are targetted with unsolicited emails, perhaps relating to sexual matters or supposed investment opportunities.

The extensive use of **CCTV technology** means that an individual may be caught on TV around 10 times a day. CCTV can be used to prevent and detect crime. However, it could also be used to track the movements of innocent individuals; this information, in the wrong hands, could be used against the interests of the individual.

Legal issues ◄

The growth of ICT technology has lead to new fields of criminal activity; it also provides new ways of carrying out crimes such as fraud.

case study 2
▶ **illegal network access**

(This is from an article in The Times in August 2007.)

A man who was seen using his laptop in the street has been arrested on suspicion of illegally logging on to a wireless broadband connection.

Two officers saw a man sitting on a garden wall outside a home in West London. When questioned, he admitted using the homeowner's unsecured broadband connection from his position on the wall. He was arrested under the Computer Misuse Act and Section 125 of the Communications Act 2003.

The first conviction for this offence was in 2005. A man from Ealing is believed to have been seen sitting in his car using his laptop outside the house of a Wi-Fi subscriber in West London. He had been seen in the neighbourhood before by a local resident and reported to the police. When police examined his laptop, they discovered that he had logged on several times before. He was found guilty of dishonestly obtaining an electronic communication service. His computer and wireless card were confiscated, he was fined £500 and given a 12-month conditional discharge.

1. Under which section of the Computer Misuse Act would the man have been arrested?

2. What should the householders have done to prevent the illegal access?

3. What equipment was needed to make use of the Wi-Fi connection in the street?

ICT technology can be used to encourage terrorism. Sites appear on the Web that contain instruction manuals about guns and explosives as well as inciting users to violence.

The Internet is used by paedophiles who establish websites where photographs are shared and contacts maintained. Children can be groomed by paedophiles that join chat rooms pretending to be children themselves. After establishing Internet contact with a child over a period of time, the paedophile might arrange a face-to-face meeting. A police report states that online gaming and social-networking websites are being frequented by sex offenders seeking vulnerable children.

A number of websites have been set up that sell prescription drugs, such as Viagra and steroids, at high prices. These drugs are not easily available in shops. In 2007, a man was sentenced to four and a half years in prison for running a counterfeit drug-selling business on the Internet. He purchased the fake Viagra and other popular drugs, which can usually only be obtained in the UK on prescription from a doctor, from factories in India, Pakistan and China. The counterfeit Viagra tablets were packaged to look like the real thing and were sent to customers who had ordered Viagra through websites.

Some fictitious sites have been created, often to look like well-known sites, where users are encouraged to purchase goods that are never delivered or to part with personal information including credit card details.

Activity 3

1. Prepare a leaflet for children on the dangers of using the Internet
2. Research examples of recent convictions for Internet misuse and present your findings as a written report or slide show.

Social issues ◄

The growth of ICT technology means that an employee can keep in touch with his work colleagues, where ever he might be. With many organisations working globally, an individual may find that he is on call for work 24 hours a day, seven days a week as he needs to keep in contact with people in different time zones. Stock traders may need to check the markets in different parts of the world at any time of the day or night. Such demands can adversely affect an individual's ability to relax and lead a life away from work.

Web sites such as YouTube are being misused by some users, often children, to bully others. This has become known as **cyber-bullying**. Mobile phone videos are loaded on to websites that show fellow pupils and teachers being attacked or humiliated.

Computerised stock control systems have allowed supermarkets to reduce costs by reducing stock levels and selling a greater range of products. This has resulted in the supermarkets being able to push down prices; many local small shops have lost trade as they have no longer been able to compete with nearby supermarkets and have been forced to close down. This means that consumers have reduced choice, no longer having the convenience of a nearby shop.

Computer systems can be very expensive to develop and install. A corner shop that installed a very sophisticated, bar-coded POS system found that the benefits that followed did not make up for the very large installation cost as there was no scope for reducing staffing.

Consequences for individuals and for society ◄

ICT has already had a great effect on society, for example, in the fields of employment, leisure and communications. This trend is likely to increase as processors get faster, computer memory gets bigger and the price becomes comparatively cheaper. In this section, we look at some of the ways ICT has affected the way we live and work.

▶ ICT and education

Computers are commonly used by pupils and students in schools, colleges and universities to research topics and produce word-processed reports. The Web has become a major research tool. Most students at university and college are provided with free Internet access. This means that students can access material, provided by their tutors and from a wider base, at any time. This makes them more independent and less reliant on their tutors. There is also less pressure on the library for books when all the students on a course are preparing for the same assignment as the students can access material stored on the Web. If they have to miss lectures, they can access the notes online. The tutors can contact students using email which is quicker, more reliable and cheaper than sending out printed notes. Students can submit word-processed work attached to an email; this is less likely to get lost and the time of submission is recorded.

ICT has improved education in other ways. Intelligent tutoring systems enable the computer to give the student information, ask questions, record scores and work at the individual's pace.

Computer communications provide new opportunities for distance learning. Students can send their work to their tutors by email and receive back annotations and comments.

Videoconferencing may be used for lectures, enabling two-way communication and discussion. This is particularly useful in remote and under-populated areas. Students in schools or colleges can undertake courses which are not viable to run in their own institution as there are not enough students wishing to study the subject. From a college viewpoint, the use of distance learning increases the potential market for their courses.

Computer-based training (CBT) is a sophisticated way of learning with help from ICT. A simulated aircraft cockpit used for pilot training is an example of CBT which is cost-effective and less dangerous than the real thing. The number of computer-based learning packages that are available is growing very quickly as modern computers have the processing power and storage capacity to support fast-moving and realistic graphics.

Many schools are using electronic means for registering their students in class. Such methods can provide up-to-date attendance information brought together from a number of classes. This provides a valuable overview of an individual student's attendance as well as providing the opportunity of studying trends in class attendance.

case study 3
► studying online

Figure 24.1 Jane

Jane works in Africa as the pharmacist in an Aids research project team. For her work, she needed to extend her knowledge of certain drugs. There were no suitable courses available to her locally. She could have flown to the US or Europe to attend a course. This would have meant that she would have had to take several weeks away from work and would have cost the charity for which she worked a considerable amount of money.

Instead, Jane enrolled on an online course run by a university in the United States. The course was designed to take 60 hours of study. The university had run the course delivered in the traditional way before expanding to include online delivery. The online delivery had expanded the potential market for their course, as any English speaker with access to the Internet could apply, and had provided an increase in income for the university. Once the course was set up and all the materials were produced, the course cost less to run online than by using traditional methods.

Jane was able to study in the evening at home after work; she linked to the Internet and accessed her learning materials online. The course was divided into sections and at the end of each section there were questions to check her understanding and progress.

Jane was assigned a tutor. She communicated with him through email, attaching her word-processed assignments when they were completed. Of course, she was not able to have face-to-face support from her tutor.

Online study can cause a student to feel isolated as they do not meet and work with fellow students. To help get around this problem, several online discussion groups were set up so that students could share ideas and obtain help and support from each other. As the students on the course were spread around the world, Jane had to choose a group whose members would be awake when she was!

Jane successfully completed the course and acquired the extra knowledge that she needed to carry out her role in the research team.

1. List other ways that Jane could have acquired the knowledge that she needed.

2. Explain why online learning was the best option for her.

3. Identify drawbacks of online learning.

Schools and colleges enter students for public examinations electronically using **electronic data interchange (EDI)**. Files are sent to the exam boards with details of students and the modules that they are entering. Later, the results are sent back to the institution, in encrypted form, via the Internet where they are imported into a local database. The results are then printed out for the students.

Students applying to university through UCAS do so using the **Electronic Application System (EAS)**. The student fills in an online form, entering personal details, course choice details, lists of qualifications gained and examinations to be taken, and a personal statement. Once this has been completed, the student's reference is added by a teacher. The application is then sent electronically to UCAS which forwards it to the chosen institutions.

Activity 4

Copy and complete the grid below, identifying 10 ways in which ICT is used in your school or college. For each describe the benefits and limitations of using ICT.

Use of ICT	Benefits	Limitations

▶ ICT and commerce

Business today is unrecognisable compared with business in the days before computers. The **electronic office** is an obvious example of the effect of ICT on business. A modern office is likely to have a computer on every desk. The work carried out in offices is generally the receipt, processing, storage and dispatch of information. The computer can do all these things more efficiently than traditional methods. The use of word-processing and desktop publishing packages has meant that even very small businesses can produce material in a very professional manner and so enhance their image with both existing and potential customers. A document can be stored for later use and does not need to be retyped if it is needed again, perhaps with minor changes.

In most large shops, an electronic point of sale (EPOS) system is used. This has a scanner that is used to read the bar code that identifies the product. The software uses functions to look up the price and description of the product and produces an itemised receipt for the customer. This lists the description

Figure 24.2 The chip-and-PIN card reader in a supermarket

of every item purchased, together with its price and the overall total that the customer has spent. The software also maintains sales and stock level information for use by the store manager. By linking to an **electronic funds-transfer (EFT)** system, an EFTPOS till allows customers to purchase goods using credit or debit cards, as their credit worthiness can be checked online. "Chip and PIN" cards provide security for the customer, who has to enter a secret four-digit code (see Figure 24.2).

The growth of the Internet has provided organisations with many opportunities. Products can be ordered online using the Internet. This can make the potential market for a product very large – even worldwide. Money can be transferred electronically to pay for goods and services.

On the other hand, an over-reliance on ICT can cause some problems; when for some reason the technology is not available, perhaps through breakdown, normal business cannot continue. Many large stores, such as supermarkets, now use only EPOS tills that are linked to a central computer that holds a database of stock levels and prices. There have been many cases when a store has had to ask its customers to leave the store because a computer failure has made all the tills inoperable.

The storage of large amounts of data can result in information being accessed inappropriately.

Shopping via the Internet (**e-shopping**) is taking an increasing share of the market. It is not just hi-tech and multinational companies, such as Amazon, that use e-commerce to sell goods. Businesses can set up their own website to market their products. The site can be used to make customers aware of special offers. A family-owned butcher from Yorkshire, Jack Scaife, started to use an Internet site (http://www.jackscaife.co.uk) to sell its bacon, sending deliveries all over the world. Soon e-commerce was bringing in £200,000 worth of sales from a site that cost only £250 a year to maintain.

E-shopping lets the shopper make purchases without having to leave home. It allows access to products that might be hard to find in the local area. Those who have difficulty getting to shops, due to poor health, family commitments or time constraints are enabled by e-shopping to make the purchases they desire. Sellers are no longer limited to a local pool of potential purchasers and do not have to maintain the overhead of mailing expensive catalogues needed for mail order shopping.

Some people are put off e-shopping because of worries over security issues relating to the use of credit and debit cards over the Internet. Others do not like to choose items without seeing them physically.

case study 4
► the cashless society

There have been many developments in ICT that are leading to a society without cash. These include:

■ credit cards, where computers store financial details

■ cheques, which are processed by computers using magnetic ink character recognition (MICR)

■ direct debits, which are used to pay regular bills, generated by computer

■ wages and salaries, which are paid by electronic transfer

■ phonecards, which are used to pay for telephone calls

■ smart cards, which are used for automatic debit and credit

■ electronic funds transfer, which connects a shop with the banks' computers.

1. Many people welcome the fact that they do not have to carry around large amounts of currency. Why do you think this is so?

2. Other members of society feel 'left out' of the move towards a cashless society as they cannot have, or do not want to have, credit and debit cards. Identify the categories of the population who might feel this way and give reasons why they might do so.

3. Make a list of all the transactions that you think will continue to require cash in the future; for example, giving pocket money to young children.

4. Describe the benefits and drawbacks of a cashless society. When preparing your answer consider different categories of people – the young, those on a low income, the elderly.

Activity 5

Explore four different e-shopping sites on the Web. Try a variety of sites – include a small specialist site as well as a large food retailer.

1. Describe the common features that you find.
2. A member of your family is trying to decide whether or not to become an e-shopper. To make sure that they make a reasoned decision, list three benefits of e-shopping and three drawbacks.

► ICT in the home

Computers have changed many people's leisure activities; a high percentage of households now have a home computer or a games console. These offer a new form of entertainment. For example, many home users use software to trace and record their family tree, to plan a journey or to access information from an encyclopaedia held on a CD-ROM. Digital photography, with photographs taken using digital cameras, stored and printed in the home is increasing in popularity.

Home use of the Internet is also growing. The use of email to keep in touch with family and friends around the world is becoming increasingly popular. Home users can also book holidays, carry out their personal banking and order goods that will be delivered to their door from a supermarket. Such e-commerce facilities save time and people can carry out these tasks at a time that is suitable to them.

Activity 6

Carry out a survey of 10 home computer users who have access to the Internet. Find out how many hours a week they spend on the Internet. List the different features and services they use. Enter your results into a spreadsheet and display them graphically.

Access to the Internet also provides leisure opportunities. Development of chat rooms, applications such as Friends Re-United and online games provide many opportunities to the home user. Of course, there are dangers as well as benefits from such a wealth of facilities. Many parents worry that their children spend so much time sitting at a computer or games console that their health may be damaged; there are many sites, such as those displaying pornography, that are unsuitable for children.

► Changes in employment patterns

Technological developments since the Industrial Revolution over 200 years ago have led to changing patterns of employment. The Industrial Revolution led to the building of factories that resulted in a shift in the population from the countryside to the towns. The development of the computer has also changed patterns of work and it has affected nearly every part of industry and commerce.

Some skills have disappeared completely. For example, in the printing industry, typesetting used to be a skilled operation using hot metal. It was performed by print workers who had undergone a seven-year apprenticeship. Now it can be done in the office using a desktop-publishing program and a standard PC. This has resulted in greater job flexibility and the breakdown of the traditional demarcation lines between printers and journalists. Football-pools coupon checkers are no longer needed as the job can be done automatically.

Some jobs, such as gardener or delivery driver, may have changed little but they may be affected by such inventions as computer-controlled greenhouses or automated stock control. Other jobs, such as supermarket checkout operator, bank clerk or secretary, have changed considerably. This has usually meant that existing staff have had to retrain to use ICT.

Tele-working

Tele-working means using ICT to work from home. It has been made possible by advances in technology and networking such as fibre optics, faster modems, fax, palm-top computers, satellite systems, internal email and tele-conferencing. Many organisations have a **Virtual Private Network (VPN)** that enables workers to access the network from a home computer via a broadband connection or from a laptop at a WiFi hotspot. Working from home has been common for a long time in some fields, such as sales representatives, telephone sales and the self-employed. Now tele-working has extended home working to other fields.

It is now possible for workers such authors, journalists, computer programmers, accountants and word-processor operators to do their work by tele-working. This means that they have no need to travel, which can result in less stress, saving time and money.

The Britannia Building Society has implemented a new policy for its text-creation department – what used to be called the typing pool. Dictation of letters is now done over the phone, stored and then transmitted to the tele-worker's home. The completed documents are typed at home and submitted by email. Britannia say that the system works really well and it is easy to monitor the work rate and error rate of the tele-workers. The typists can work flexible hours to fit around family commitments

It is common for tele-workers to spend part of their working time at the office and part at home. Typically such a person would work at home for three days and be at the office for two. Many companies use a method of "**hot-desking**", where, instead of having a workstation for every employee, tele-workers share the use of a number of computers. For the employer, tele-working saves the cost of office provision such as desks, chairs, floor space, heating and car parking space.

The pool of available labour for a job is hugely increased if tele-workers are employed. An employee is able to move away from a crowded city and relocate in an environment of their choice which may provide cheaper housing and a better quality of life. British Airways flight booking takes place in Bombay, proving that you don't even need to be in the same continent. Much computer programming and testing for British companies is done in Asia.

There are obvious advantages in tele-working for the employee such as flexible working hours, avoiding time-wasting rush-hour travel and the associated costs. Some childcare problems can be eased. Tele-workers are no longer tied to living in crowded cities where housing costs are very high but can live in the location of their choice.

However, the lack of the social side to work can be a disadvantage for employees as many people make friends through their work. Not everyone is good at getting down to work by themselves and some find it hard to be part of a team at such a long distance. Much informal training takes place in a workplace without the participants really knowing about it. There is a real danger for many people that the distinction between work and private life gets blurred and work takes up more and more of their time, affecting the quality of life and producing stress. Tele-working demands new skills, and training is essential if it is to be successful.

In the long term, if tele-working were to become the way the majority of people worked there would be implications for society. Reduced travel would be environmentally friendly, could cut pollution and, perhaps, could lead to the end of cities and offices as we know them.

case study 5
▶ tele-working in action

Figure 24.3 Rosemary

Rosemary works as an events organiser for a London museum based in the east of London; she lives in the west of the city. Travelling to work involves Rosemary taking three trains and walking for a total of 20 minutes. On a good day, she can complete the journey in under an hour and 15 minutes but she often meets delays that can add up to an hour to her journey time. She finds the travelling tiring and stressful and resents wasting so much time in unproductive activity. When she made the journey on a daily basis, if she awoke feeling slightly unwell she would have to stay at home and miss a day's work as she did not feel that she could cope with the travel.

Rosemary negotiated with the museum to become a tele-worker for two days a week. Now she stays at home on a Tuesday and a Wednesday. She is able to plan, produce reports and contact external providers without interruption from her colleagues. She has a home computer with Internet access that she uses to keep in touch with her colleagues via email.

Rosemary makes good use of the travel hours saved on a Tuesday and Wednesday. She usually gets up at her normal time and either goes for a morning run when she would otherwise be sitting on a train or else starts work early and then meets up with a friend for a long lunch break.

The balance of working three days at the museum and two days at home suits Rosemary. She would not like to work at home every day as she would miss interacting with her colleagues.

1. Summarise the benefits of tele-working to Rosemary.

2. List any benefits to the museum.

3. List five jobs where tele-working would be appropriate, explaining why.

4. List five jobs where tele-working would not be appropriate, explaining why.

Activity 7

Complete the following tables, including as many points as you can:

Benefits of tele-working

To the employee:	
To the employer:	
To society:	

Limitations of tele-working

To the employee:	
To the employer:	
To society:	

Problems in the workplace

Ready access to ICT in the workplace can cause problems. Some employers have found that they need to control their employees' access to social-networking sites during work time as many working hours are being lost. A college has blocked access to E-Bay during the core working hours as many students were using computers meant for private study to access the Web site.

► ICT and banking

The number of transactions carried out by banks has grown so rapidly that they could not now operate without computers. Banks transfer money electronically. Most workers are paid directly into their bank account by computer. Many regular payments such as mortgages, utilities and insurance premiums are transferred electronically from an individual's bank account to the company's bank account as a direct debit or a standing order. This means that an individual does not have to worry about remembering to pay bills, risking extra charges if they forget to pay.

Cash is not as important as it used to be (see Case Study 4). Most individuals do not need to carry as much cash as before thanks to credit and debit cards. The bank keeps detailed records of purchases made by a customer using their card. A customer is sent a list of all transactions every month which helps them in their financial planning. These records can be used by customers to show proof of purchase of a product if they have not kept the receipt and wish to return the product.

If a person does want cash, they can visit an **automated teller machine (ATM)** at any time. Before banks introduced ATMs, they had to employ more people to act as tellers who worked at a desk and dispensed money to customers. The customers would have to queue up until a teller was free, fill in a withdrawal form and show identification before they were given cash by the teller.

The development of ATMs has brought benefits to both customers and banks. Customers now have access to cash at any time of the day or night, which is much more convenient. It saves time for the customer who no longer needs to queue at the bank counter behind someone who is depositing 20 bags of small coins! The bank reduces its need for staff or can use its tellers for other tasks. It can close some branches and provide an ATM device, which reduces costs. Many banks have used the space and staffing freed up by the installation of ATMs to develop new services which are useful to customers and increase profits.

There are also drawbacks to the use of ATMs. There has been a growth in fraud and card theft as well as attacks on customers who have just withdrawn money. Many customers preferred the more personal approach when human tellers were used. There have also been occasions when computer failure has occurred putting many ATMs out of action.

Home or online banking means that users with access to the Internet can check their bank account balances, transfer money between accounts and pay bills from home. This can be much more convenient than having to visit a bank in person and transactions can be carried out more quickly than by using the normal postal service. Online banking has the added convenience of its services being available at any time from anywhere in the world. However, many people are unwilling to use online banking as they are afraid of fraud and other security issues.

ICT has also affected our shopping habits in other ways. The widespread use of credit cards means it is possible to shop and pay for goods without leaving home. Cable and digital TV shopping channels and the Internet have provided new ways of finding out what to buy instead of the traditional catalogues.

▶ ICT and medicine

Computers are used in the administration of hospitals and doctors' surgeries, storing patients' records. Pharmacists keep records of customers and their prescriptions. When a patient's records are needed by a health professional they are always available, unlike paper-based records that can only be viewed in one place at a time. The use of electronic records can lead to concerns about the consequences of storing inaccurate data and increased threats to the security of data.

case study 6
▶ NHS "Choose and Book"

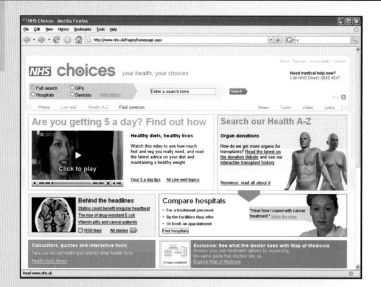

Figure 24.4 NHS website

An Internet booking system for NHS services, called "Choose and Book", is a new national service that aims to provide an electronic booking system with a choice of time, date and place for a patient's first appointment as an outpatient. This new system is designed to give patients a much greater involvement in the making of decisions about their treatment

Patients can choose from one of four or five hospitals and information on these hospitals should be available to General Practice doctors and other staff, as well as patients on http://www.nhs.uk. Patients have the choice of booking their hospital appointment electronically when they are at the doctor's surgery with the help of the GP, or they can book later by themselves using the Internet.

1. Find out more about the new system on http://www.chooseandbook.nhs.uk.

2. Explain the main benefits of the "Choose and Book" system.

3. Imagine that you were in charge of the team who were implementing this new system. What would be your main actions to ensure that the system worked as intended?

Some hospitals are now experimenting with storing medical records on smart cards kept by the patient and taken with them every time they visit a doctor, dentist, pharmacist or hospital. The smart card can store a complete medical history and can be updated at the end of each visit.

ICT also helps in the diagnosis and monitoring of patients' illnesses. **Expert systems** can be set up to help in diagnosis by asking questions about symptoms and using the answers to draw conclusions. Computer-controlled, ultra-sound scanners enable doctors to screen patients very accurately. X-ray film is being replaced by on-screen digital pictures. Computers can be used for continuous monitoring of patients' bodily functions such as blood pressure, pulse and respiration rates. Such systems provide instant feedback of information and can free up nurses to carry out other duties.

SUMMARY

The use of ICT is influenced by the following factors:

▶ cultural

▶ economic

▶ environmental

▶ ethical

▶ legal

▶ social.

In banks, computers are replacing human workers and most people are paid directly into their bank account by computer. An increasing number of other transactions take place electronically. Some people predict a cashless society.

ICT is widely used in medicine:

▶ Much administrative work is performed by computer.

▶ Expert systems are used as an aid to diagnosis.

▶ Computers are used to monitor the bodily functions of patients.

ICT is playing a growing role in leisure in the home, where the Internet is used by many people who explore the Web and send emails to friends and family.

In education, ICT supports online learning. CBT is used extensively in training. The Internet and encyclopaedia CD- and DVD-ROMs are used for research purposes. ICT plays a major role in educational administration.

Tele-working is the name given to the use of ICT to work from home.

ICT brings many advantages but there are also drawbacks that must be acknowledged and understood.

Questions

◀

1. Describe three different ways in which a company might make use of the Internet to benefit its business. (6)

Jan 2005 AQA ICT1

2. Describe two ways in which the Internet has helped the authors in writing this book. (6)

3. Many people are now working from home.
   a) Describe the benefits to the employee and the employer of tele-working. (8)
   b) Describe the disadvantages of tele-working. (8)

4. "The digital divide is turning our country and the world into information haves and have nots". Discuss the consequences of this digital divide. (14)
   In this question you will be marked on your ability to use good English, to organise information clearly and to use specialist vocabulary where appropriate.

5. Discuss the benefits and limitations of online shopping to retailers and their customers. (9)

AQA Sample Paper

Index